SOTHIS PRESS

Sothis Press aims to completely reimagine the public domain book with an emphasis on universal access to captivatingly designed print editions. We believe that technology will enable every single book ever published to be perpetually accessible in beautiful print form. Sothis is beauty in ubiquity.

www.sothis.press
1 Waterhouse Square, London, ECIN 2ST

The Pasteurization of Milk From the Practical Viewpoint

A Handbook Relating to the Installation, Operation and Control of Pasteurizing Plants, Useful to Milk Dealers, Students in Dairy Schools, Public Officials Having Control of Milk Handling, and of Interest to the General Public

Charles Herbert Kilbourne

SOTHIS PRESS

ISBN 978-1-0368-0620-0

The Pasteurization of Milk From the Practical Viewpoint is in the public domain.

This edition is published by Sothis Press © 2023. The company is registered in England and Wales. The Company Number is 14900389.

The

PASTEURIZATION OF MILK

FROM THE PRACTICAL VIEWPOINT

A HANDBOOK RELATING TO THE INSTALLATION, OPERATION AND CONTROL OF PASTEURIZING PLANTS. USEFUL TO MILK DEALERS, STUDENTS IN DAIRY SCHOOLS, PUBLIC OFFICIALS HAVING CONTROL OF MILK HANDLING, AND OF INTEREST TO THE GENERAL PUBLIC

BY

CHAS. H. KILBOURNE

LATE CHIEF OF THE DIVISION OF PASTEURIZING PLANTS
NEW YORK CITY DEPARTMENT OF HEALTH

NEW YORK
JOHN WILEY & SONS, INC.
LONDON: CHAPMAN & HALL, LIMITED
1916

INTRODUCTION

In presenting this book on the pasteurization of milk, the author has endeavored to so handle the subject that the milk dealers, the students in dairy schools, and health department officials who are charged with the supervision and control of the milk sold in their communities, will find in the book practical assistance in their work. Except for the excellent bulletins issued by the United States Government, there are few books which deal with the practical aspects of milk pasteurization, and the author felt that a real need existed for such a work.

It is possible that his long association with the milk work, as it has been carried on by the New York City Department of Health, has somewhat colored his views with official conservatism. He has tried, however, to be fair and impartial, and to look at matters from the dealers' point of view. He has presented freely practical lessons which he has learned during a somewhat wide experience in investigating pasteurizing plants.

If any dealers with questionable motives look in this book for practical suggestions by which they may be able to save milk and cream which is going bad, or if they wish to learn how departmental regulations can be evaded, they will be disappointed.

It is hoped, however, that dealers who honestly desire to handle milk so that it will be safe and salable will find assistance in this book.

The author wishes to acknowledge the assistance and good advice which he has received from Dr. Wm. H. Park, Dr. Chas. F. Boldune, Ole Salthe and James E. Thomson, and from Inspectors Ryan, Lee and others in the New York City Department of Health. Also he wishes to thank the various manufacturers of dairy machinery for information given and illustrations supplied.

C. H. K.

New York, September 1, 1916.

The Pasteurization of Milk from the Practical Viewpoint

CHAPTER I

PASTEURIZATION IN GENERAL

The guarding of food supplies in large towns and cities is one of the most important functions of government activity. This activity, like all forms of organized effort, has been a gradual development. In the most primitive communities, individuals naturally felt a certain care over the food which they ate. This interest was instinctive and fundamental, since each individual was conscious that some foods were dangerous. He therefore obeyed the law of self-preservation when he selected his food.

As the homes of individuals became more congested on account of the increase in population, it grew to be increasingly difficult for the individual to select the food which he ate. He therefore came to depend more and more upon the judgment of other people who had either used similar food before or else knew how it was produced. The substitution of the judgment of others for his own personal inspection of food naturally developed into a certain sort of communal or public control of foods. As communities became more and more congested, it became necessary to centralize the control of foods under some definite organization. This centralized control, which was usually at first vested in some private organization, was later taken over by the government which obtained in the town or city. Thus we have the Boards of Food Control, which are now usually centered in the Health Departments.

As the inter-dependence of the various towns and cities became more fully recognized, it became evident that not only the cities but the states should on broad lines, regulate the production and sale of foods. This state regulation of foods has been centered partially in State Health Departments, partly in the Agricultural Departments and sometimes in special departments whose duties have been to control the foods sold in the state. These state departments are becoming increasingly efficient as the good work which they do is becoming more and more recognized.

Since the means of communication and transportation are becoming more rapid, the effect is that the world is becoming smaller, and so the consumption of foods is not confined to the states where it is produced. The increasing use of cold storage facilities for preserving foods also adds to the wide area from which cities and towns can obtain the foods which are there consumed. It, therefore, becomes evident that state regulation is too narrow. In order to properly protect the citizens, therefore, the general government has taken up, to a large extent, the control of foods which enter into interstate commerce, and government standards are adopted.

Among the articles of food which have perhaps received the largest share of attention we find milk. This attention has been deserved on account of the almost universal use of milk, and also on account of the fact that it becomes easily infected with dangerous bacteria, and having become infected it is itself an ideally good medium for their growth. The milk thus becomes so changed or decomposed that it is frequently unsafe for food. The opportunities for infection are many, on account of the many persons through whose hands it passes before it reaches the city consumer, and where it is perhaps fed to the children who depend upon milk for their life. At any point along the line of travel from the cow to the consumer it may become dangerously infected. If the cows from which it is drawn are not healthy, the milk may start on its journey in an unsafe condition. Even if it is perfectly wholesome when drawn from the cows, it may at any point along its line of travel become changed from its natural state.

Disease-producing bacteria or other less harmful germs may enter the milk from stable dust or from the manure and urine which drops from the cow herself. The milk bottles and cans may have been carelessly washed or the water with which the washing is done may be from contaminated sources. Diseased, or unclean persons may be employed in milking the cows, or in handling the milk after it is drawn. Cans or bottles in which the milk is placed may have been returned from the consumer in filthy condition and have been again used for containing fresh milk before being thoroughly washed and sterilized. Contagious disease may have been in the homes from which they are returned.

Much of the milk in large cities is handled at

creameries or shipping stations which are located in the country and from which the milk is shipped to the cities. Thus, there is an added opportunity for milk infection on account of the insanitary buildings or from coming in contact with dust-laden air or diseased employees. Once having gained access to the milk the growth and multiplication of the germs is aided by the powerful agencies of age and heat. If germs which produce disease get into the milk the danger that epidemics of disease will occur is a serious one. Each point in the handling of milk is therefore an important point, for the chain of milk safety is only as strong as its weakest link.

Aside from the danger of bacterial contamination there is of course the possibility of actual adulteration being practised within the legal definition of that term. Foreign substances may be added, such as water, coloring matter, preservatives, etc., or the cream may be removed. Such adulteration may be accidental or it may be deliberately practised, with "malice aforethought," with intent to deceive and defraud the final consumer, for the purpose of increasing the profits. On account of the sharp competition which exists in modern trade, some dealers consider that they are justified in doing some things which are objected to by public boards of control. Sometimes milkmen will endeavor to save a perishable product by the use of preservatives. This form of adulteration, however, is at present very little practised.

The importance which attaches to the public or communal control of milk is not the only force which is lifting to a higher level the character of the milk sold to the public. Milk dealers themselves are realizing more and more that modern conditions are such that only the best and the most efficiently produced and handled product can remain in the market. They are therefore becoming more and more active in establishing means for effectively controlling their own supplies. The methods employed are various. Associations of men have been formed, the object of which is to secure mutually useful advice and to give each other the benefits of individual experiences.

The larger dealers establish within their own organizations an efficient inspection service and seek by means of laboratory tests and research work to study their problems and so secure the best milk possible at the least expense, and with the least amount of waste. Thus the governmental forces and the dealers themselves both work toward the goal of a better milk supply.

The great importance of the problems which affect the securing of a safe milk supply has led to the publication of many books, in which the varions phases of the matter have been considered. The composition of the milk, its chemical characteristics, the number and kind of bacteria contained, the forms of adulteration which occur and the methods for detecting such adulteration have all been broadly covered. The relation of milk to the public health and the disease epidemics which have been due to infected milk have received large attention in published works. Much has also been published regarding the most effective methods to be employed in the public control of milk supply.

While there is an extensive literature concerning the problems of milk pasteurization dealing with the matter from the theoretical side, and also covering certain special phases of the work, it is rather surprising to find how little of this is available for the practical use of the milkman who wishes to install pasteurizing apparatus and to so handle his output that the most efficient service can be secured with the least possible expense. This dearth of literature has been no doubt largely due to the fact that the information which the many practical men have gained in their experiments has never been so brought together that it is available for the use of those who are seeking knowledge along practical lines. That this should be so is but natural, for several reasons.

First: The manufacturers who are interested in emphasizing the good points of their apparatus are not inclined to call attention to their defects.

Second: Milk dealers who, through experiments in trying out various forms of apparatus have learned what appears to be best suited to their needs, are not inclined to publish to the world information which has cost them much time and money to secure. Should they do so, they would be benefitting their competitors.

Third: Public officials in city, state or national public service, who have made many inspections, conducted tests and carried on extensive research work are comparatively few. In fact, in very few cities is the field for observation sufficiently extensive to enable such officials to draw unbiased or broad conclusions. State officials are as yet doing little work along these lines. The general government has been and is now carrying on extensive and valuable experimental work, and as a result much useful and valuable information has been published.*

The writer has published some matter which his experience has taught him.† Dr. Rosenau has devoted considerable space to the subject of milk pasteurization and has given much valuable information on the subject.‡

Additional information which is of considerable value has been acquired by the experience of the writer and other inspectors in the New York City Department of Health, working in conjunction with the laboratories. It seems to be proper that this information could be available to the general public.

As before stated, the application of heat to perishable foods in order to preserve them from de-

*Ayers and Johnson, Bureau of Animal Industry—Circular 184 and other publications.

†N.Y.C. Department of Health Reprint Series Nos. 1 and 27.

‡The Milk Question by M. J. Rosenau.

cay has probably been practised since the discovery of fire. The earliest record of the application of heat to food in a scientific way was in 1872, when Sheele, the well-known Swedish chemist, discovered that vinegar could be preserved against spoiling by heating it. Soon after this other people began to preserve fruits, vegetables, milk, etc., by heating them in closed vessels. No scientific explanation, however, of the reason why the heating process preserves these foods appeared until 1860 to 1870, when Pasteur, in his studies as to the causes for the deterioration which took place in wines and beers, discovered that there was certain forms of life previously unknown whose unchecked growth gave to liquor its bad flavors, etc. These were termed by Pasteur "diseases." He discovered that by heating beer and wine to a temperature of 158° to 176° F., and holding the fluids at these temperatures for a period of time, the "disease germs," as he called them, were destroyed and the flavor of the liquor was improved, the loss by spoiling being at the same time prevented.

This discovery was of immense financial benefit to Europe, and in consequence the heating of wines and beers became very generally practised, the process being termed "pasteurization," from the name of the man who originated it. At the present time, no brewery is considered to be well equipped and up to date unless extensive apparatus is employed to heat the product before it is placed upon the market.

The thought that the same principle might be applied to the preservation of milk on an extensive scale had its birth in Europe, and it is claimed that the first commercial milk pasteurizer was made by Ahlborn in Hildeshein, Germany, in 1880, and that a similar apparatus was made in the same year by Fresca in Berlin, Germany. The process was later adopted in Denmark, where dairying has reached such a high degree of perfection.

At all events, the machine now known as the Danish pasteurizer was introduced into this country by Reid about 1895, and is now sold under various names by different manufacturers. Modifications and developments of the original idea form the basis of nearly all the pasteurizing machines used at the present time.

This form is illustrated in Figs. 1 and 2, and is described more in detail later in this book.

In Denmark the object sought by the use of the

pasteurizer was the improvement of the flavor and quality of butter, the manufacture of which is such an important industry in that country. The pasteurization of the cream used for butter is there generally practised, and it is subsequently ripened by the introduction of a pure culture grown in the laboratories in order to secure the most desirable texture and flavor. These methods have resulted in the production of a butter in Denmark which is considered to be a standard article. The industry has grown remarkably and its products are shipped to England and other parts of Europe in normal times, and have also, in times of shortage, been shipped to the United States.

When it became appreciated in Denmark that the spread of bovine tuberculosis was a dangerous menace to the dairy industry, a law was passed requiring that milk whey, etc., which was fed to calves, swine and other animals must be heated before being fed in order to prevent the spread of this disease among the growing stock. This proved so useful in accomplishing its object, that similar laws have been passed in various states in this country. Vermont and Pennsylvania have such laws in the East. Strange to say, many of the dairymen object to this, much preferring the unheated skim milk. The dangers are not generally appreciated.

It became apparent to some people that that which was good for preventing disease among cattle would also be useful in preserving health among children, and in 1866 Soxhlet advocated that milk which was fed to infants be heated. About this time, Jacobi, the distinguished American pediatrist, advised the boiling of milk fed to infants. In 1869 he referred to the apparatus devised by Soxhlet, this being the first American reference to the subject. This was before it was so well known, as is now generally accepted, that milk is a means of carrying infectious diseases, such as diphtheria, typhoid, scarlet fever, etc. Jacobi found that children fed upon boiled milk were less liable to have intestinal disorders, and his long and continued success in treating children tends to discredit the assertions made by some that the pasteurization of milk at comparatively low temperatures destroys its nutritive value and is injurious to children. Jacobi still advocates the boiling of milk.

It was soon brought to the attention of milk

dealers that milk which was old and liable to become a commercial loss on account of souring could be saved if it were heated, and that the fermentation due to the growth of the enormous number of bacteria in milk could thus be checked. Milk which had been handled in an unclean manner, or which had been badly cooled by insufficient icing, or which it became necessary to hold for a long time before sale, could still be made marketable by rapidly heating and cooling it. Little attention, however, was paid to the matter of a uniform heating of the milk, or to the exact degree of temperature to which it was subjected. Almost no attention was directed to the destruction of pathogenic bacteria. These, in fact, were little understood by the milk dealers, and few dealers maintained laboratories for studying milk problems.

On account of the questionable motives which actuated dealers in adopting this method of milk treatment, as well as the faulty methods employed, health authorities, as a rule, at first looked upon the pasteurization of milk with disfavor. It was considered that the process was a makeshift and simply used as a means of covering up bad practices in the production and handling of milk. Instead of encouraging the pasturization of milk, or even ignoring it, health authorities rather took steps to discourage its practice. The desire to secure a safe milk supply for the public took form in strenuous efforts to clean up the sources from which the milk was secured, to secure proper refrigeration and to assure clean handling. The cry was for clean milk as produced from the dairies, and the process of pasteurization was frowned upon.

In New York and other cities, commendable progress was made in the effort to secure a clean, safe raw milk, and the improvement made in the conditions under which milk is handled, particularly in receiving stations and creameries, can hardly be appreciated by those who were unfamiliar with conditions before the campaign for clean milk was inaugurated. Dairy conditions have also been greatly improved, and the dairymen themselves repeatedly say that they would not for any consideration go back to old conditions. All efforts made to advance the production of clean milk should be continued with unabated vigor.

It soon became apparent, however, that the securing for a large city of a raw milk supply which could be guaranteed to be reasonably clean and safe was practically impossible. If it was secured, it would involve the expenditure of an enormous amount of money and energy, which the taxpayers would not approve, and it was seen that it might result in so much added expense to producers and dealers that the price of milk to the consumer would be greatly increased, perhaps to such an extent that the sale of this valuable universal food would be seriously curtailed.

Public health administrators were at length led to the conclusion that the pasteurization of milk was the only practical means of securing a supply which would be reasonably safeguarded against dangerous infection. By many persons it was considered to be a makeshift, and it is still so regarded by some authorities, it being held by them that its employment should only be continued until a clean, safe raw supply can be secured. Others, and by far the majority of authorities, are, however, convinced that pasteurization will always be a necessity, and that even the best of raw milk should be provided with the additional safeguard which will be secured by pasteurization.

It is even urged by some that certified milk,

which is the safest raw milk possible to be obtained, should be pasteurized before being used for infant feeding. This thought has been strengthened by information of comparatively recent discovery, that milk may become infected through coming in contact with persons who harbor disease germs, but who are not in any way ill. These persons are therefore perfectly innocent of any knowledge that they are a source of infection to milk, and thus a dangerous and insidious menace to the milk consumer. It has been found that certain persons may be, and not infrequently are, infected with typhoid bacilli, without they themselves knowing that they are thus infected. Such persons are known as typhoid carriers. Other persons harbor the germs of diphtheria in mouth and throat, without having any symptoms of the disease whatever. Such persons may handle milk and unwittingly infect it.

Within the last few years there have been in different cities epidemics of septic sore throat. These have occurred in Boston, Baltimore, Chicago, etc. In Boston in 1911 several thousand persons were made ill by milk coming from a farm where the milk was exceptionally well cared for.* Many of the persons thus affected died. In 1912 a similar epidemic occurred in Baltimore, where many persons were made sick and many deaths occurred. This was traced to milk infection.† Had these milks been properly pasteurized, these epidemics of disease would probably have been avoided.

Aside from the foregoing considerations, the fact that a very considerable percentage of cows producing milk are tuberculous gives an added reason why it is desirable to heat the milk before use. There has been much misapprehension in the public mind regarding the actual danger which threatens human beings from this source. There seems to be no doubt that these dangers, while real, have been somewhat exaggerated. In the first place, the fact that a cow may be seriously affected with tuberculosis does not make it certain that her milk is dangerous to use. If proper care is exercised while the milk is being drawn from the cow, none of the germs will get

*Winslow, C. E. A., Journal of Infectious Diseases, Volume X No. 1, January, 1912.

†Public Health Reports, November 22, 1912.

into the milk. Of course, if the disease existed in the udder of the cow, or in the milk ducts, there would then be direct contamination of the milk as it was drawn. Cows more frequently, however, have the disease in the lungs or other parts of the body, in which case the germs will pass through the alimentary tract and be voided with the urine and fæces. If the cows are allowed to become badly soiled, particles of manure and dirt, carrying with them the tubercle bacilli, will inevitably fall into the milk, thus infecting it. It is also possible that dried manure in stables where tubercular cattle are kept will produce dust, which, when stirred up by the movements of the cattle or by persons and by sweeping, may fall into the milk. This danger is probably more imaginary than real, since recent experiments by the Geneva Experiment Station * indicate that stable dust plays a part in milk infection which is of comparatively little importance. It is also true that the exposure of manure to light and air for a length of time sufficient to produce a dry dust would tend to destroy the tubercle bacilli. The amount of infec-

*Bul. 409. N. Y. Ag. Exp. Sta.

tion would be therefore comparatively small. It is also a known fact that while tubercle bacilli will live in milk, they will not multiply.

The apprehension experienced by many persons that they may be infected with tuberculosis from the milk which they consume should be lessened by their knowledge of this further fact, namely, that tuberculosis of the bovine type with which cattle are affected is rarely infectious to human beings who are over fifteen years of age, and rarely causes fatal tuberculosis in children over five years old. The careful examination of hundreds of cases where death has occurred from tuberculosis among children appears to conclusively show that the danger of infection from bovine tuberculosis is largely limited to persons under fifteen years of age. Of deaths from tuberculous affections of persons under this age, only 10 per cent. had tuberculosis of the bovine type. This has been determined for New York City by investigations carried on by Park and his colleagues, extending over several years, and the same conclusions have been reached by European scientists.

The records of New York City for the years

1913-1914-1915 show the following interesting facts:

DEATHS OF CHILDREN UNDER FIFTEEN YEARS.

	1913	1914	1915
From all causes...	23,574	22,312	23,031
From all tubercular diseases	1,288	1,166	1,200
10 per cent of these latter figures indicate	128.8	116.6	120.

These represent the number of persons who died in these years from tuberculosis which was probably of bovine type.

This means that out of every 1,000 deaths from all causes of children under fifteen years of age, 5.4 in 1913, 4.8 in 1914, and 5.2 in 1915, had tuberculosis which was probably of the bovine type.

While it is possible that bovine tuberculosis may reach human beings through infected meat or by direct infection from dust, dirt, etc., it is probable that most of it is due to infected milk.

There are many cases of tuberculosis existing

in bones and glands which do not result fatally. Nearly one-half of these are thought to be of the bovine type, and very likely have their origin in milk.

It is evident, therefore, that while the danger of tubercular infection from milk when expressed in terms of percentages is comparatively slight, there appears to be, nevertheless, a definite danger to children from this source, and the safeguarding of this milk by pasteurization is extremely desirable.

Rosenau * places tuberculosis first among the diseases carried by milk, thus considering it a greater menace than typhoid fever.

All these facts have induced most health authorities to change their attitude toward the pasteurizing of milk for general use, and it is now the exception instead of the rule to find persons familiar with the subject who favor the use of raw milk, except when it is from dairies which are absolutely above suspicion.

The important consideration which influences authorities when they insist that milk be pasteurized is of course the desire that the milk may be free from dangerous bacteria.

*The Milk Question—M. J. Rosenau.

There are commercial reasons which influence dealers, such as the desire to prevent the souring, and thus the loss of milk, but these considerations have little weight with health authorities. The prevention of economic waste of food products has a bearing upon the public health, and there are indications that the personal interest which those in authority have as men in these matters is being extended to their official interest. Care must be observed, however, that these two considerations be justly balanced.

The United States Government has endeavored to reduce the losses which result from the condemnation of cream which is not considered fit to use as cream. This is accomplished by partially denaturing it by the addition of salt, so that it can be used for butter making. An attempt has also been made to prevent the destruction of milk which has a temperature above that fixed by the authorities as legal. It is well known that in some cities this drastic action is taken in order to secure proper refrigeration. Such destruction of milk, which has a high food value, should be looked upon as almost *criminal.*

Careful experiments carried on by Park, Rosenau, and others have demonstrated the fact that if milk is heated to a temperature of 140° F., and is held at this temperature for twenty minutes, or even less, all of the pathogenic or disease-producing bacteria will be destroyed. Also, if the milk is heated to a uniform temperature of 158° F., and held thereat for one minute, all such disease germs will be killed. Between these extremes, different degrees of temperature, and correspondingly varying lengths of holding time, will produce equally good results. These experiments have been carried on of course in laboratories where ideal conditions for securing accurate results existed. Accurate temperatures could be obtained and the length of time for which the milk was held could be carefully controlled. When, however, pasteurization is carried on under commercial conditions, where large quantities of milk are handled, the same amount of accuracy is not possible. There are variations in temperature and variations in the length of time for which it is held, due largely to the mechanical defects or limitations in all forms of machinery. These factors of error render the results uncertain, and it is therefore desirable that the temperatures and the lengths of time for holding, which in the laboratory tests have been found to be effective, be somewhat increased, in commercial practice.

Various cities have made regulations in this respect which appear to best suit local conditions. In New York City, when regulations were first adopted concerning milk pasteurization, the length of time for which milk must be held when heated to 158° F. was officially fixed at no less than three minutes, and if heated to 140° it must be held for at least twenty minutes. Between these extremes of temperature holding times were fixed which varied between three minutes and twenty minutes.

Experiments were made by Schores and Rosenau * in which careful tests were carried on while milk was being pasteurized in the ordinary way, a commercial pasteurizer and holding device being used. These tests were made for the purpose of determining if the thermal death point of pathogenic bacteria, as indicated by laboratory tests, could be safely relied upon under commercial conditions. In their experiments the investigators introduced germs of tuberculosis, diphtheria, and typhoid fever, into milk, which was then run

*The Journal of Medical Research Vol. XXVI No. 1.

through a commercial pasteurizer and heated to different temperatures, and held for varying lengths of time. The results obtained indicated that when an attempt was made to heat this milk in a commercial pasteurizer to a temperature of 140° F., and then hold it for twenty minutes, there was a considerable unavoidable variation in the actual temperature to which the machine was capable of heating the milk. This was due to the difficulties of both hand and automatic temperature control. Even with the best automatic controller, it is probable that the real variation of temperature is greater than is indicated upon the record charts. Thermometers are not usually sufficiently sensitive to record rapid fluctuations in temperature. It is also true that thermometers indicate *average* temperatures only, and it is possible that certain portions of the milk, the temperature of which is considerably below this indicated average, pass through the apparatus.

In regard to the time held also, especially when a continuous or flow type holding apparatus is used, great uncertainty exists. It is impossible to be sure that the time is the same as intended. The rate of flow varies, and there is also more or less mixing of the milk due to various causes, which will be discussed later.

Schores and Rosenau found that pathogenic bacteria were not all killed when the factors of temperature and time were attempted which had been found effective in laboratory tests. They therefore recommended, in order to provide a sufficient margin of safety, that all milk he heated to a temperature of at least 145°, and there held for no less than twenty-five minutes. This opinion is shared by most persons who have made a study of the subject. The New York City Department of Health, in 1914, so modified its regulations as to require that pasteurized milk be heated to from 142° to 145° F. for at least thirty minutes.

While the destruction of all pathogenic bacteria in milk is the primary concern of health authorities, there are certain other considerations which have weight with them, and among these is the effect which heat has upon the chemical characteristics of the milk. It has been found that when milk is heated to say 160° F. or over, some of the constituent parts of the milk are changed in such a way as to make it somewhat inferior as a food, especially for infants. These changes have been somewhat extensively studied by the United States Government experts, and it has been found that as the temperature rises above 160° F., these changes take place to an increasing extent, and at the boiling point quite extensive changes are observed. At this temperature, the milk sugar becomes scorched, the casein and the albumin are somewhat hardened, the calcium, magnesium and phosphoric salts are partially precipitated, all of which renders the milk less easily digested. It therefore follows that the lowest temperature which is consistent with the destruction of the pathogenic bacteria is the most desirable.

It is also true that milk which contains an excessive number of bacteria carries with it, as a result of their growth, certain by-products, such as acids and other undesirable substances, and may also contain dangerous poisons. Rosenau states that the known poisons which are produced by the growth of bacteria in milk are very few, but that such poisons may possibly exist and produce harmful effects upon the persons using the milk. These by-products of bacterial growth in milk are not entirely destroyed by heating, even when the boiling temperature is attained. For this reason it is desirable that milk which is to be pasteurized shall, before it is heated, be as free as possible from all bacteria. It is this consideration which has led health authorities to insist that the milk before pasteurization shall not contain an excessive number of bacteria.

Milk dealers know that their customers commonly judge of the richness of the milk by the amount of cream which appears upon the top of the bottle. For this reason the showing of a "good cream line" is one of the things for which the milk dealer strives. In some parts of Europe the cream is thoroughly mixed in the milk by homogenizing it, so that the cream will not rise, but the American housewife is not educated to this custom, and therefore anything which interferes with the apparent volume of cream on the milk is a stumbling-block to the milk dealer. It has been found that milk which is heated to a temperature of 148° F. or over and is held for any considerable period of time, will have its cream line destroyed, or at least be made indistinct. When milk is heated to a temperature lower than this and is held for a great length of time, the same effect is produced. There are other reasons which interfere with the cream line which will be discussed later.

Another objection which customers urge against pasteurized milk is that it has a distinctly scorched or cooked taste. This is undoubtedly true when the milk is heated to a high temperature, or when an uneven temperature is maintained. If the heating medium is kept at a point greatly above the temperature to which the milk is to be raised, the same effect is produced. On account of this cooked taste it is evident that there is a commercial objection to pasteurization where it is not properly or carefully done.

Moreover, when a low temperature of heating is used, the enzymes are not much affected, while they are destroyed when a high temperature is applied. Indeed, the Storch test, which is used to determine whether skim milk has been heated to 165° or higher in compliance with the laws in this regard, depends upon the non-action of the enzymes. These are destroyed by high heating. This test is of little or no value in testing milk pasteurized at low temperature.

Since an even heating at comparatively low temperatures and a holding for a long time secures the desired result in the destruction of bacteria, and at the same time produces none of the undesirable effects which follow when a high temperature is employed, it is evident that this method of procedure is that which is much to be preferred.

To summarize the considerations which have just been discussed, it is evident that the pasturized milk which most nearly fulfills the requirements desired by health authorities, and at the same presents, from the commercial aspect, the least objections, must conform to the following demands:

First: Before pasteurization it must contain only a small number of bacteria.

Second: When it is heated it must be subjected to a temperature which destroys all pathogenic or disease-producing bacteria, and also a large percentage of bacteria of all other kinds.

Third: Its constituents must be very little changed or altered by excessive heating. Its nutritive value must not be impaired, and its digestive enzymes must be uninjured.

Fourth: The taste of the milk must not be noticeably changed.

Fifth: The apparent volume of cream must not be reduced.

In so far as these factors represent results which are properly demanded by health authorities, there are, broadly speaking, two general methods by which they can be controlled when their control depends upon the powers exercised by the municipality or the state. One of these may be called the *laboratory method,* and the other the *inspectional method.* When the *laboratory method* is employed it is necessary that certain definite chemical and bacteriological standards be adopted, which apply to pasteurized milk or to milk which is to be pasteurized. Some of these standards may depend upon color reactions. Such standards being in force, the effectiveness of control depends upon the frequent examination of samples of milk taken before and during the process of pasteurization and as it is offered for sale. If the tests show that the milk conforms to the legal standards, it may be judged, with a fair degree of confidence, that the apparatus employed in pasteurizing and handling the milk is properly installed and operated.

When the *inspectional method* of control is employed, it is necessary that certain definite regulations be adopted which shall specify on broad lines the form of apparatus which must be used, how it shall be installed and how it shall be operated. The construction and arrangement of the buildings and of the various rooms may be specified, it having been determined by previous experience that such and such apparatus and surroundings operated and handled in such and such a way will produce milk which will be of the character desired. With these regulations in force, it is possible, by frequent and careful inspections made of pasteurizing plants, to determine whether or not the legal regulations are being faithfully observed. If it is found that they are complied with, then it is logical and safe to conclude that the milk produced is of suitable character, and will satisfactorily respond to all suitable chemical and bacteriological tests which may be applied.

As a matter of fact, neither method of control is used by authorities to the exclusion of the other, but a combination of the two methods is found to be the most satisfactory. In some cities the emphasis is placed upon laboratory tests, with inspections made to confirm the results. In other places, the emphasis is upon physical regulations and inspections, with occasional laboratory tests to confirm the conclusions reached and to aid in the discovery of obscure defects which may escape observation during the usual inspections. No advice can be given as to which method is most desirable without first knowing local conditions, taking into account the extent of the milk industry, the available factors of men, money and laboratory equipment, etc. Each community presents its individual problem.

Health authorities are often requested by milk dealers to give advice as to the form of pasteurizing apparatus which will perform the best work, or they are asked what kind of machine the authorities recommend. It is manifestly improper for health authorities to recommend specifically any one or more forms of apparatus as the best. If they did, they would lay themselves open to the charge of unfair favoritism, and to the suspicion that improper influence determined their judgment. It is only proper that they should state that in their experience certain forms of apparatus, operated in certain ways, have performed certain things in actual practice which are or are not satisfactory to those who are responsible for this condition of the milk. Indeed, when advice is asked as to the form of apparatus most desirable to be installed at a given point, it is impossible to solve such a problem without possessing complete knowledge of the conditions which surround each plant where it is desired to install a pasteurizer. Each plant presents an individual problem of its own, which must be carefully studied, having clearly in mind all the facts which specially relate to the plant in question. It is impossible to place too much emphasis upon this point, since time and money and the securing of expert advice before the construction and equipment of a plant will frequently bring continuous dividends in convenience, in the saving of time, in preventing needless waste of money, and in the general satisfaction which accompanies the successful operation of an efficient plant.

Among the factors to be considered, the construction and physical characteristics of the building where the plant is to be located is of importance in determining the selection of the apparatus to be used. Again, the distance of the plant from the point where the milk is to be distributed and consumed should be considered. If located in the country at a distance from the city, the element of time available after the milk is delivered by the dairyman is also an important item. The amount of milk to be pasturized daily, and the use which is to be made of the finished product have a bearing upon the final decision. With many dealers the question of expense may be of first importance, but it must be borne in mind that cheap installation does not always mean economy. The question of expense involves both the initial cost of installation and the outlay for operation and upkeep. Where the matter of furnishing sufficient steam is a point for serious consideration, or where water expense affects the final cost, it is desirable to use those forms of apparatus which consume the fewest heat units and which use the least amount of water. There are also other considerations which will be discussed later, and it will be found that at many plants unique conditions exist which must be carefully studied.

On broad lines, however, it may be stated that a pasteurizing plant, in order to be as nearly as possible ideal in regard to its construction, and which will at the same time be capable of producing a milk which will be satisfactory from the standpoint of safety and wholesomeness, and will also satisfy commercial demands, should combine the following features:

1. *The Building.*—The building where the plant is located should contain ample space and should, if possible, be so arranged that the milk will flow by gravity from the point of receipt, or where cans are emptied to the point where it is taken out for delivery or transportation.

2. *Separate Rooms.*—The various stages in the process of treating the milk, such as receiving, pasteurizing, cooling, filling containers, storage and of washing the containers, should be carried on in separate, clean, well-lighted and ventilated rooms or compartments.

3. *Protection from Contamination.*—At all points during the process of handling, the milk should be protected against contamination from contact with unclean apparatus, and from infection by means of dust-laden air. This protection is especially necessary after the holding process is completed. Any infection occurring after this time is not corrected by heating, and may therefore be a source of danger.

4. *Constant Temperature.*—The heating appa-

ratus, whatever its form, should be capable of heating every particle of milk to a temperature which will never be below that to which it is the intention that it shall be heated. For instance, if it is intended to heat to 140°, the temperature should not fluctuate between 135° and 145°, but should remain constantly at 140° or over.

5. *Method of Heating.*—The heating should be accomplished by an apparatus which allows the milk to flow over a clean metal surface, which is kept hot by means of heated water, and not by direct steam contact. This hot water should never have a temperature more than 10° or 15° higher than that to which it is desired to heat the milk.

6. *Holding Apparatus.*—The holding apparatus, whether the absolute or the continuous flow type process is employed, should be so constructed and operated that it will definitely insure the holding of every particle of the milk without fail for the desired length of time.

7. *Recorders.*—The temperature to which the milk is heated must be automatically recorded, and should be automatically controlled. It is desirable that a temperature recording device be installed at the point where the milk leaves the

heating apparatus and that another recorder be attached to the outlet pipe from the holder. This latter recorder indicates whether or not any great amount of heat has been lost during the holding process, and also shows whether the apparatus is thoroughly heated before the milk is admitted. If not so heated, the first milk leaving the holder will have been somewhat cooled, and the recorder will indicate this. After the apparatus has been in operation for a considerable time, it will then acquire the temperature of the milk, which will then pass out with the temperature but little diminished. These observations do not apply to a tank or batch pasteurizer. In this case a single recorder attached to the tank itself shows all that is necessary. It also indicates the time for which the milk is held at the required temperature.

8. *Pumps and Pipings.*—The amount of milk piping in use should be reduced to the minimum requirements. All piping which is used should be of sanitary construction, with the joints and turns so arranged that they can be easily taken apart for cleaning. As before stated, the apparatus should be so arranged that the milk will flow by gravity from point to point in the process of

treatment. If, however, it is necessary to use milk pumps, they should be of absolutely sanitary construction. The use of pumps for moving the raw milk is less objectionable than is their use for moving pasteurized milk. Not only is the danger of infection by means of unclean pumps very serious, but if the milk is pumped while hot, the cream is so disturbed that it does not rise as readily as is desirable.

9. *Rubber Fittings and Gaskets.*—The use of rubber fittings, of gaskets and of stuffing-boxes should be avoided.

10. *Cooling.*—The milk should be immediately and rapidly cooled, to a temperature as nearly 40° F. as possible. This cooling should be accomplished within thirty minutes after the heating and holding process is completed. Except where cooling is carried on in the tanks where it is heated, the cooling can be completed in much less than thirty minutes. The milk should be, of course, kept cold until delivered to the consumer.

11. *Cleaning of Containers.*—All containers, including cans, covers and bottles, should be completely and if possible automatically washed and sterilized. They should then be thoroughly dried

and should be stored in clean, cold rooms, and there kept inverted or carefully covered until filled.

12. *Cleaning of Apparatus.*—All apparatus used in handling milk should be carefully washed and sterilized after its use, and all of it which comes in contact with the pasteurized product should be again sterilized before its use each day.

13. *Healthy Employees.*—All persons employed in handling the milk should be known to be healthy and free from any infectious disease. They should be clean in their person and habits, and should wear special clean clothing, preferably of white material, while handling milk.

14. *Bottle Capping.*—The milk bottles, after filling, should be capped by clean machinery, thus avoiding as far as possible any personal contact with the milk after it is pasteurized.

It is needless to say that all the ordinary precautions regarding the sanitary construction of buildings where food is handled should be observed, and the close proximity of stables or other contaminating influences should be avoided.

There are, of course, many other points of detail in the construction and operation of pasteurizing plants which will repay careful attention.

These will be touched upon in the following pages. On broad lines, however, it may be said that if the foregoing conditions are carefully fulfilled, then it will be possible to obtain a good safe milk which the public may consume with a feeling of security, and which should satisfy all requirements of health authorities.

It is not proper, of course, to recommend any specific type or form of apparatus to be used in pasteurizing milk. It is proposed, however, to describe in the following pages and to show by diagram or illustration the construction and operation of different makes of apparatus, and to call attention to those points in each which experience has taught are of advantage or disadvantage.

CHAPTER II

HEATERS

In 1909 the New York City Department of Health first adopted the rule which required that milk could not be considered as pasteurized unless it had been heated to a definite temperature and held at this temperature for a certain definite length of time. New York was the pioneer in requirements of this sort.

At this time various forms of milk heaters were on the market, but no device was available for holding the heated milk. All heaters employed a process which is known as the continuous or flash process of heating. By this process the milk flowed or was forced in a continuous stream through the apparatus, where it was subjected to an amount of heat which was not very definite, and which was controlled by the hand of the operator. The milk was then immediately cooled.

Most of the forms of apparatus which were then on the market are still in use, and certain new forms of heaters have been invented. Certain new types of apparatus were devised as adjuncts of the heaters, into which the milk was allowed to flow from the heaters and in which the hot milk was held for the time required by the regulations.

As before stated, the forms of heaters in use in 1909 are still employed, with certain new forms made by many different manufacturers. They are all comprised under some one of five typical forms, as follows:

First Type.—The so-called Danish heater is sold by various firms. It is shown in the accompanying illustrations (Figs. 1 and 2). This is the oldest form of heater in use in this country. In this form a large central cylinder is provided with a double cover or jacket. The space between the inner and outer shell is filled with water, into which steam is introduced. By means of a suitable device the steam is distributed through the water at various points.

In operation, the milk is allowed to flow or is pumped into the inner compartment or cylinder, and by means of rapidly revolving paddles it is thrown by centrifugal force against the inner surface of the cylinder. This surface is kept hot by means of the hot water and steam in the outer shell. The milk enters at the bottom of the cylin-

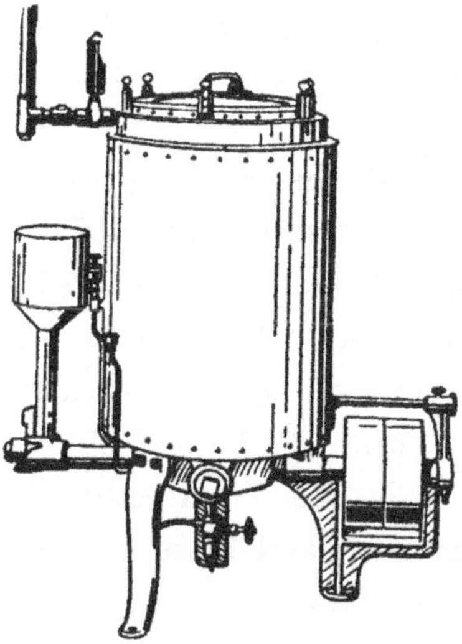

FIG. 1

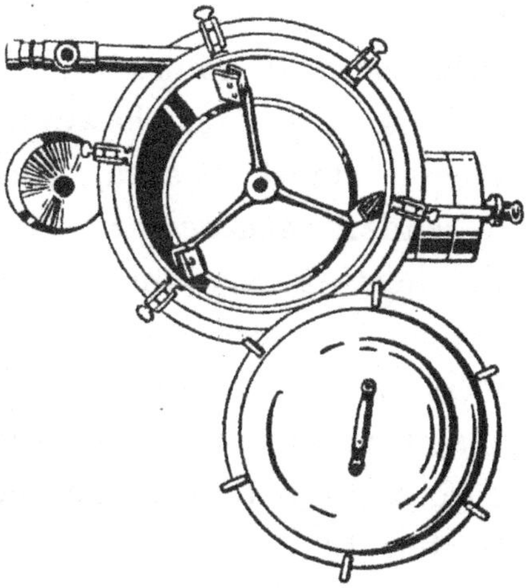

FIG. 2

der, and since the outlet pipe at the top is placed at a tangent to the cylinder, the rotating paddles tend to throw the milk out of the pipe. The machine therefore acts somewhat as a pump, and when the whole machine is set at an angle from the perpendicular, this tendency to throw the milk is enhanced and the milk can be lifted about fifteen feet. This fact is one of the advantages of the machine. Another advantage is that it is comparatively inexpensive, since it is rather small. Its size also tends toward the ease of cleaning, there being no great amount of surface with which the milk comes in contact.

This very fact, however, that the heating surface is small in area, is also a disadvantage, for it renders it necessary to raise the temperature of the heating medium to a point so high that it is frequently very much in excess of that to which it is the desire to heat the milk. The result is that certain portions of the milk reach a temperature so high that a distinctly scorched taste is imparted to it. This taste persists in the entire body of the treated milk, even though the average temperature of the milk leaving the apparatus is below 145°. This is especially noticeable if the flow of milk into the apparatus fluctuates. It can be readily seen that with a very hot heating surface, and at the same time a small amount of milk passing through the machine due to a sudden stoppage of flow, this milk would attain a high temperature.

This superheating of even a portion of the milk tends to cause an interference with the normal rising of the cream, after the milk is placed in containers. The vigorous agitation of the hot milk by means of the paddles also has a tendency to break up the fat globules into smaller particles, which do not so readily rise to the surface.

It is difficult to maintain a uniform temperature with this type of machine, since the amount of heating fluid is comparatively small, and when there is either a fluctuating flow of milk or a varying milk temperature, there is no automatic controller which can be made to respond to sudden changes in temperature quickly enough to check or increase the flow of steam into the heating chamber.

As a general proposition, also, the introduction of steam directly into a heating chamber is to be avoided, since there is always the possibility that certain portions of the steam will strike the heating surface before becoming condensed, and will therefore heat these portions to an abnormally high point.

Second Type.—There are several heating devices made by different firms, and having certain variations in the details of their construction, which embody the same general principles. These are illustrated in Figs. 3, 4, 5. In all of these the milk flows between two or more upright, heated

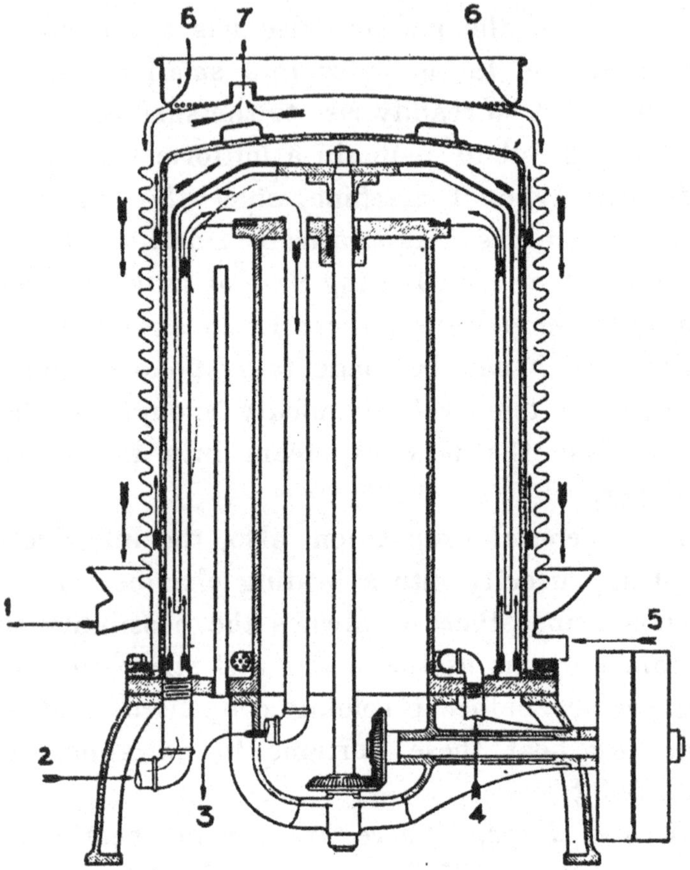

Fig. 3

1—Discharge of warmed milk to gathering tank.
2—Return of warmed milk to interior of heater.
3—Discharge of heated milk to holder.
4—Hot water intake.
5—Return of held milk to regenerator.
6—Cold milk entering heater.
7—Discharge of partly cooled milk to cooler.

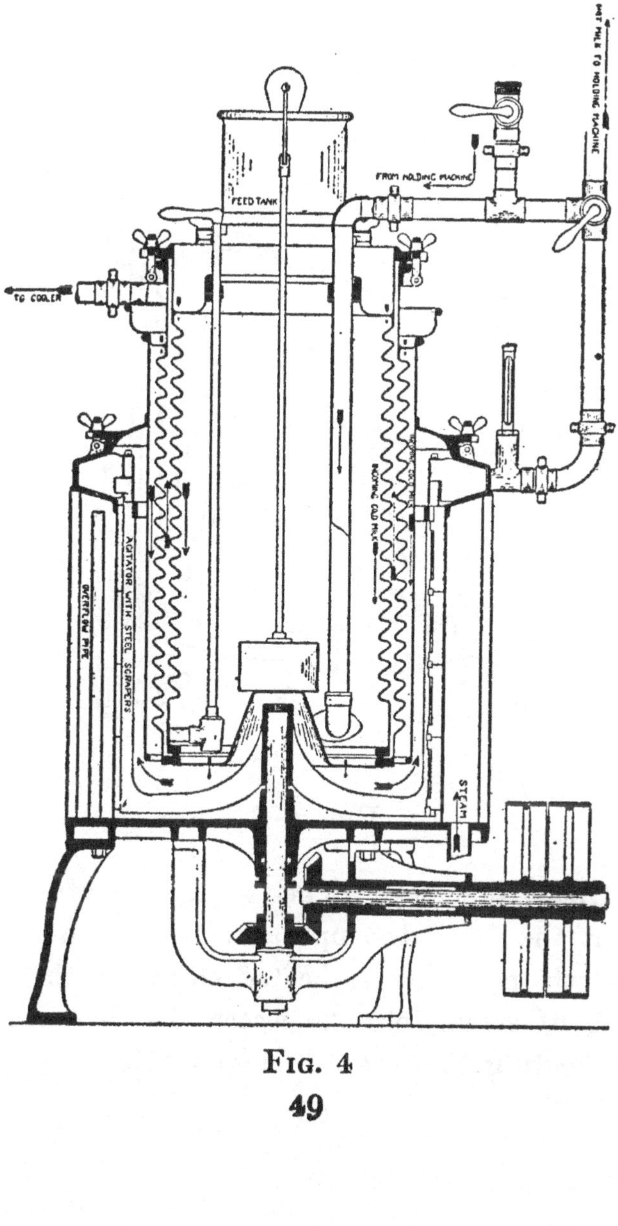
HOT MILK TO HOLDING MACHINE
FROM HOLDING MACHINE
FEED TANK
TO COOLER
AGITATOR WITH STEEL SCRAPERS
OVERFLOW PIPE
INCOMING COLD MILK
STEAM

Fig. 4

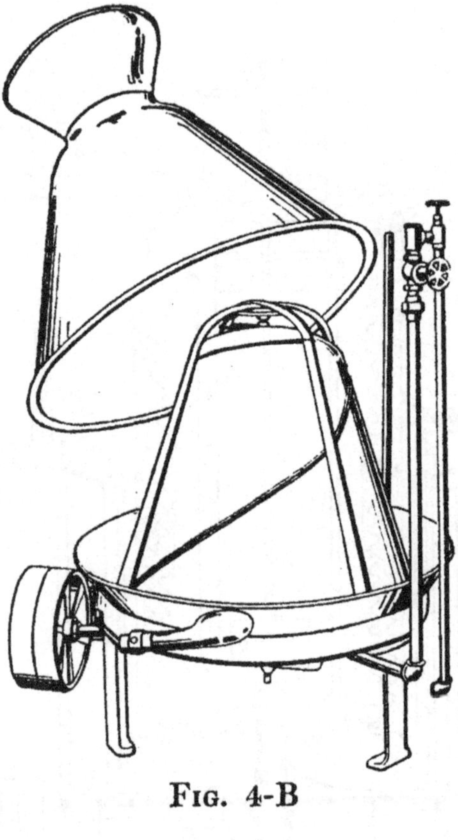

FIG. 4-B

cylindrical surfaces. The heating surfaces are kept at the desired temperature by means of hot water, which in turn is heated either by the introduction of steam into the water chamber directly, or by heating the water in tanks outside the milk

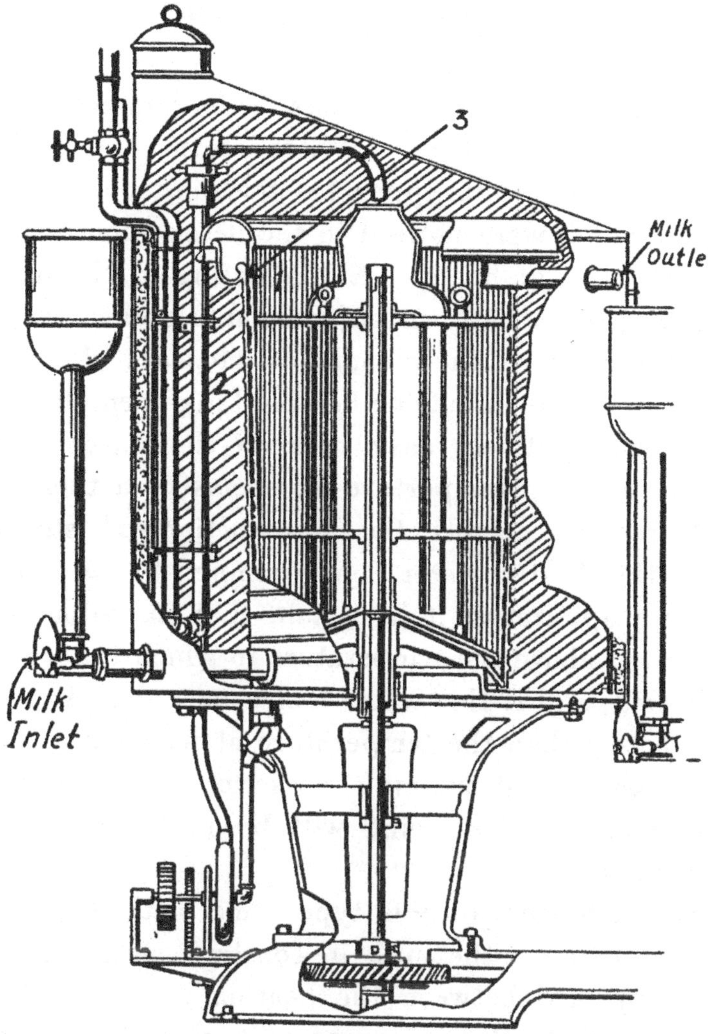

Fig. 5

1—Interior water tank.
2—Exterior water tank.
3—Space between water tanks, through which milk is forced upward.

apparatus. From these tanks it is pumped into the heating chambers. Where steam enters directly into the milk apparatus the same objections obtain which were noticed in the description of the Danish type of heater. Superheating of the milk in whole or in part, an objectionable cooked taste, interference with the cream line, etc., are likely to follow. Where an outside heating tank is employed, it is comparatively easy to keep the water at a nearly uniform temperture by means of a thermostat, which controls the introduction of steam. With this heated water being pumped to the milk-heating apparatus, it is certain that the milk can never get above the temperature to which the water is heated, and if this water is not more than 10° or 15° hotter than the temperature at which it is the intention to heat the milk, no serious superheating can result. This point will be more fully described later.

In all heaters of this type some method is employed to keep the milk in constant motion as it flows upward between the heating surfaces, thus insuring that all portions of milk come in contact with their surfaces. In the machine illustrated in Fig. 3 this agitation is accomplished by means of a revolving arm, which extends into the milk space between the inner and the outer water chambers. In Fig. 5, the inner cylinder revolves and the narrow corrugation which extends in a spiral form around the cylinder forces the milk upward. The shaft which operates these stirring devices in some instances comes in contact with the milk, and in order to prevent leakage, it is necessary to provide some sort of packing or stuffing box. This is a disadvantage, since nearly all packing is somewhat porous, and it is difficult to keep clean.

Heaters constructed in this general form have been found to be effective in operation, and good results have been obtained by their use. They have been found to be durable as well. Some individual machines have been in use for ten years, which indicates the carefulness with which they have been constructed. Those forms of heaters are the most satisfactory in which the greatest amount of heating surface is available, and where the amount of hot water which circulates through the apparatus is relatively large. These conditions render it possible to use water which has a comparatively low temperature.

Some of these heaters employ what is known

as the "regenerative principle." This means that the hot milk which flows from the machine is employed for warming the cold milk which flows into the apparatus to be heated. When this regenerative principle is used, a considerable economic saving of heat units is secured. This saving means that less steam is required for heating, and also that less ice is necessary for cooling. In the type illustrated in Fig. 4 the milk flowing from the holder is conveyed to a holding apparatus, and after being held is returned to the heater, where it flows through another compartment and helps to heat the inflowing cold milk.

At least one form of regenerative heater (Fig. 6) is so constructed that this intermediate holding process cannot be used. In sections of the country, therefore, where city regulations require that the heated milk be held for a definite period, this form of heater cannot be used. As a so-called "flash" pasteurizer, this machine has given good results.

In Fig. 4-B is shown a form of heater in which the hot water is contained on the inside of the lower cone. The milk flows from the top over the heated cone. Revolving arms to which is attached a heavy cord, keep the cone surface wiped, and prevent

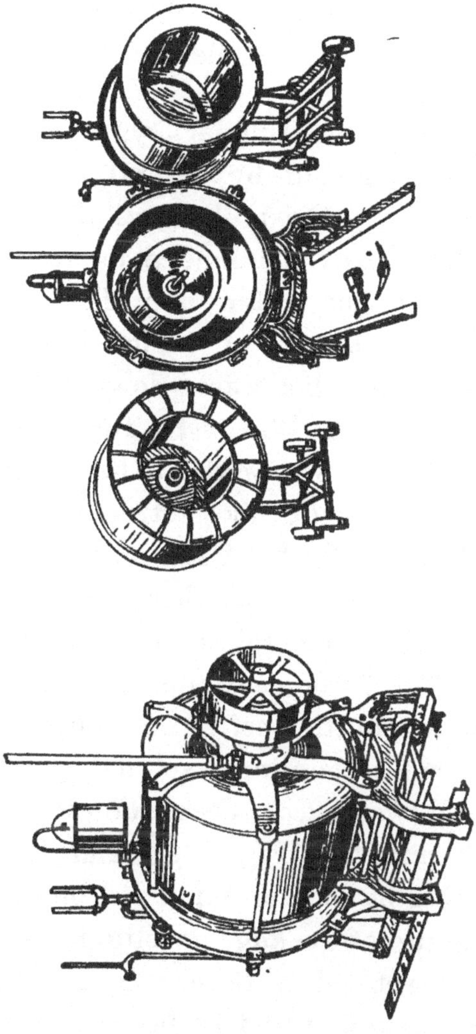

Fig. 6

the milk from sticking to the heated surface. There is difficulty in keeping the cord clean and sanitary.

Third Type.—The tubular type of heater is manufactured and sold by several concerns. In all of these forms, the milk flows through tubes, which are themselves surrounded by some heating medium, usually hot water. In the form shown in Fig. 7 the milk tube is enclosed by a larger tube, through which hot water flows in a direction opposite to the milk flow. In forms which are illustrated by Figs. 8, 9 and 10 the heating water is contained in boiler-like structures, through which the tubes extend. Care must be exercised in order to make sure that there can be no leakage between the milk tubes and the water containers. In Fig. 7 the couplings between the various milk pipes in the series are outside of and beyond the water couplings, so that no mixing of the milk and water can occur unless there is a broken milk pipe. In Figs. 8, 9 and 10 the milk tubes are either braized or expanded into the heads, and the danger of leakage is slight. Of course, when heat is applied to these tubes, the tendency is for them to expand and so become more tightly

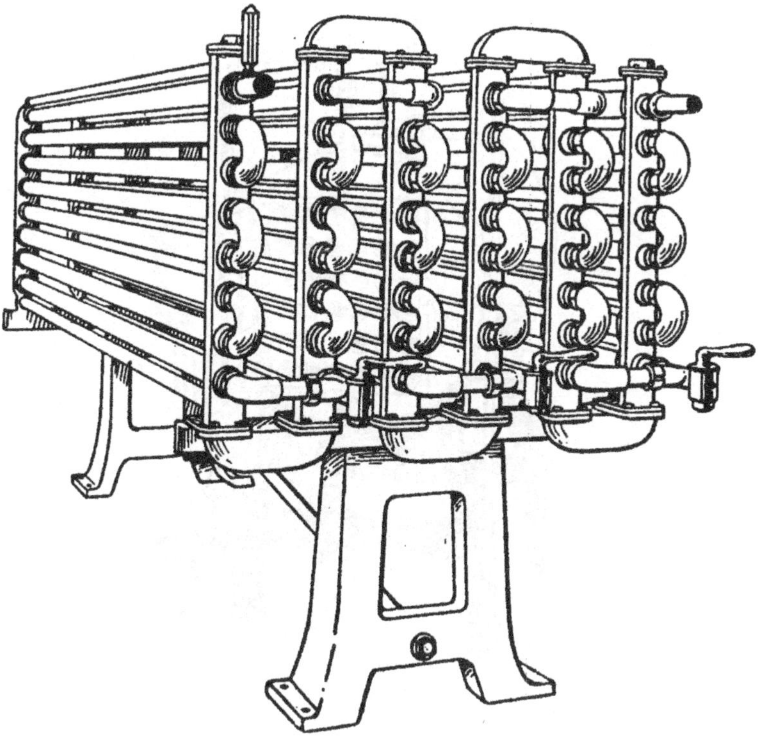

Fig. 7

fixed in the heads. Sudden and wide variations of temperature would tend to loosen the joints. The flowage of the water in a direction opposite to milk flow brings the hottest water against those milk tubes where the hottest milk is found. As the

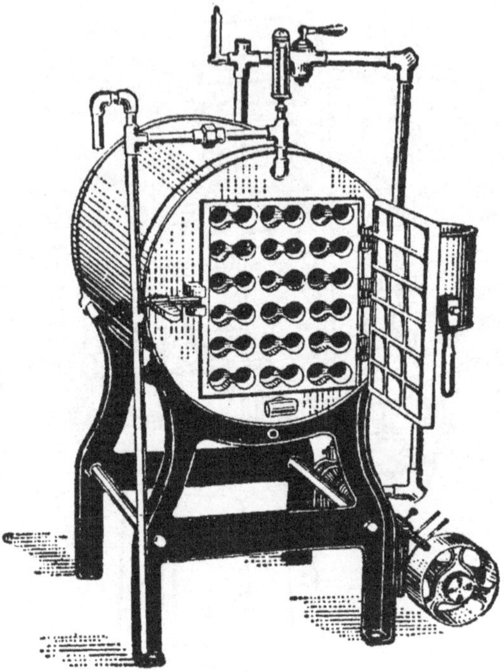

FIG. 8

water flows along the tubes it becomes gradually cooled by the cooler milk, so that it emerges at a low temperature. Thus the milk is heated gradually, and no sudden raising of the temperature occurs. This renders the control of the temperature an easy matter, since sudden fluctuation in temperature cannot readily occur. There is also

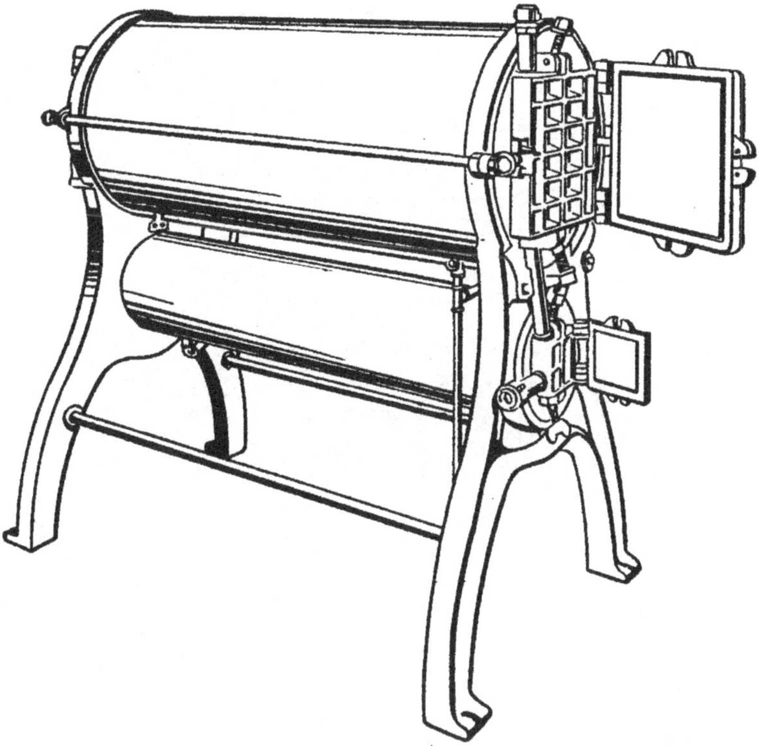

FIG. 9

an economical use of the heat, and little of it is wasted.

In Fig. 11 a form of heater is illustrated in which the milk flows through a group of small tubes, which are themselves enclosed in a large

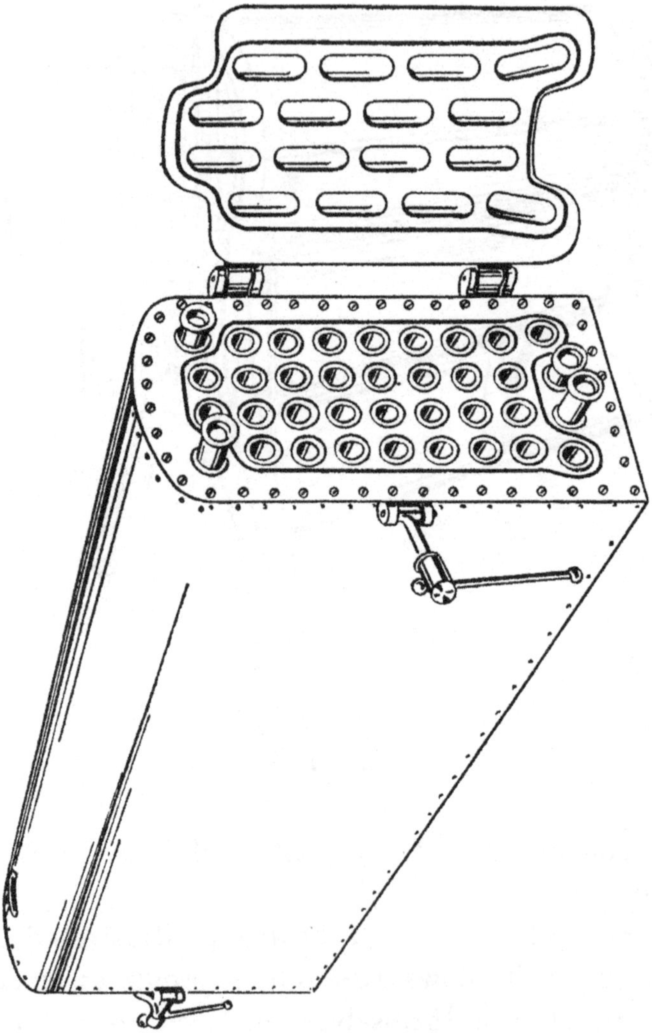

FIG. 10

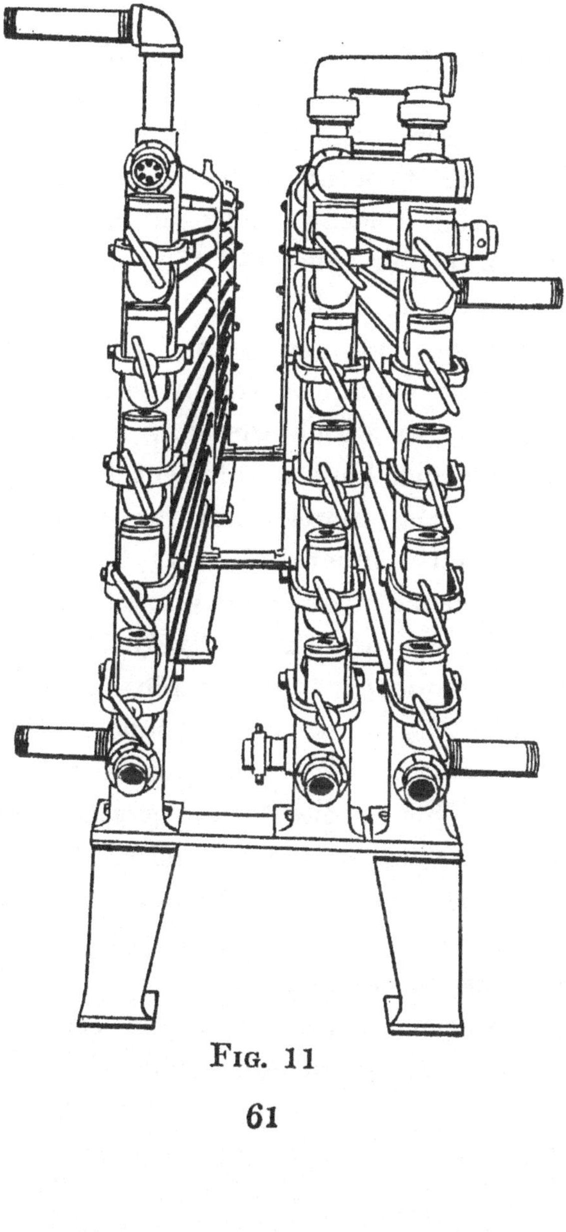

FIG. 11

tube. The larger tube contains the water which surrounds the small tubes which contain the milk. When the regenerative principle referred to is employed, in this type of heater the hot milk returning from the holding tanks is allowed to flow through the larger tube, and thus assist in heating the cold inflowing milk.

Tubular heaters are doing very satisfactory work. Those machines are in some respects the most efficient in which the flow of milk is slow enough or which have a sufficient heating surface to insure that all portions of the milk will reach the maximum temperature. It has been found from experience that when milk flows through tubes which are of comparatively large diameter, there is a tendency for that portion of the milk which is in the center of the tube to flow faster than that which is close to the inner surface of the tube. This is due to the greater friction at the point of contact. It can be readily seen that there will therefore be a tendency for the milk which flows more slowly close to the heated surface of the tube to become heated to a considerably higher point than that near the center. When the water entering the final tube of the heater is high in temperature, there is a certain amount of danger of cooking or scorching the milk, and possibly of breaking up the fat globules. It is also true that if a coating of cooked milk forms on the inner surface of the tube, it acts as an insulating material and therefore a constantly increasing temperature must be applied to the water. The return of the milk from one tube to another in the series tends to mix the milk and so break up the more rapid flowing core. This breaking up process, however, may have a detrimental effect upon the fat content and may render the cream less likely to rise. In one or two instances known to the writer it was found that when the milk left the apparatus at a temperature of only 140° the cream line was considerably interfered with.

The types shown in Figs. 7, 10 and 11 have this advantage, namely, that since they can be obtained in sections, the pasteurizing plant can be expanded by the installation of additional sections as the growth of the business demands. Since a cooling apparatus of exactly similar form is provided it is possible to extend the cooling facilities in the same way. By a proper interchange of the water used in both the heater and cooler, the regenerative principle is made use of, and the cool water leaving the heater is carried over to the cooler, so that it helps to cool the hot milk flowing into it.

In choosing a machine of this type, care should be exercised in securing one which may be easily cleaned, especially at the turns and joints. All these should be smooth and readily accessible.

It should be borne in mind that while a large heating area with many pipes is an advantage in securing an even heating with a low water temperature, the difficulties of cleaning are also increased.

It is also readily seen that when powerful pumps force the hot water rapidly through the heating pipes, a relatively lower heating temperature may be employed.

An objection to the tubular form of heater lies in the fact that the tubes in this system are nearly horizontal and parallel. There is thus but little inclination or pitch to the tubes, and the milk flows from them rather slowly when they are emptied at the close of operations for the day. This objection has been overcome in some of the more recently manufactured machines, by putting the pipes at an incline and not parallel to each other.

Thus the milk can be more readily drained from the apparatus when it is being emptied.

One striking advantage of the tubular type of apparatus is found in the fact that it may be very thoroughly sterilized. Steam can be admitted into all the parts where the milk will enter, and the steam may be under pressure if so desired.

It is important that a machine of this type be so constructed that no milk will leak from one section of tubing into another section, but that it all follows the course which it is intended that it shall follow. In the forms shown in Figs. 8, 9 and 10, the doors at the ends which cover the ends of the milk tubes are provided with gaskets to prevent this leakage.

It is well to note whether or not these gaskets are tight. In one or two forms, the gaskets consist of sheets of metal which are backed by felt or rubber in such a way that the metal readily conforms to the projections of the tubes, thus making a closely fitting cover. The metal sheet can, of course, be readily cleaned.

Fourth Type.—The tank, or so-called batch system of heating milk, is quite largely employed. Heating tanks for this purpose are now manu-

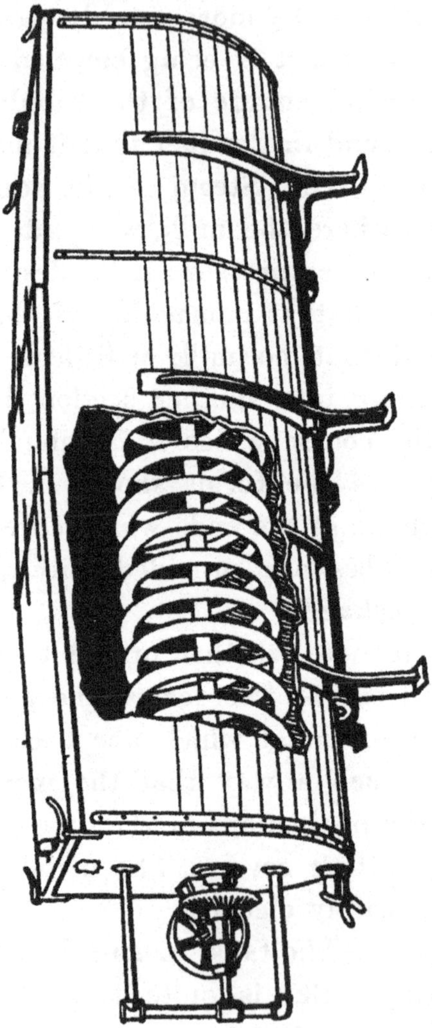

FIG. 12

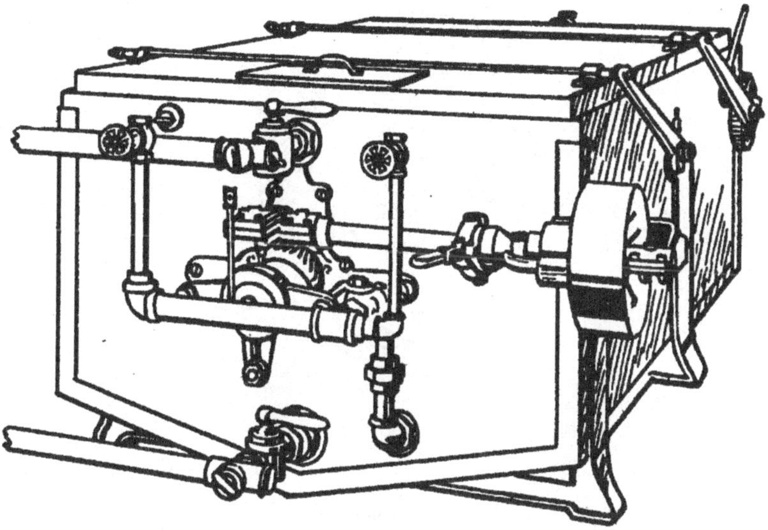

FIG. 13

factured by many concerns which handle pasteurizing apparatus. These batch pasteurizers consist, as shown in Figs. 12, 13, 14, 14-B, 15, 16, of horizontal tanks, constructed sometimes singly and sometimes made with two compartments in the same apparatus. In Fig. 16 we see a tank which is in the form of an upright cylinder. These tanks are in some cases surrounded by an insulating jacket. In others there is an open space between the inner and outer shell, through which hot water is made

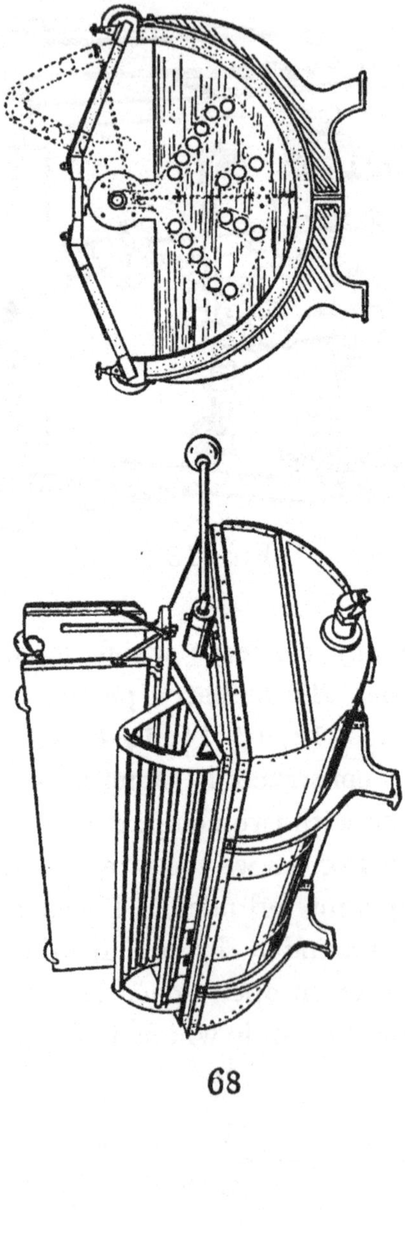

Fig. 14

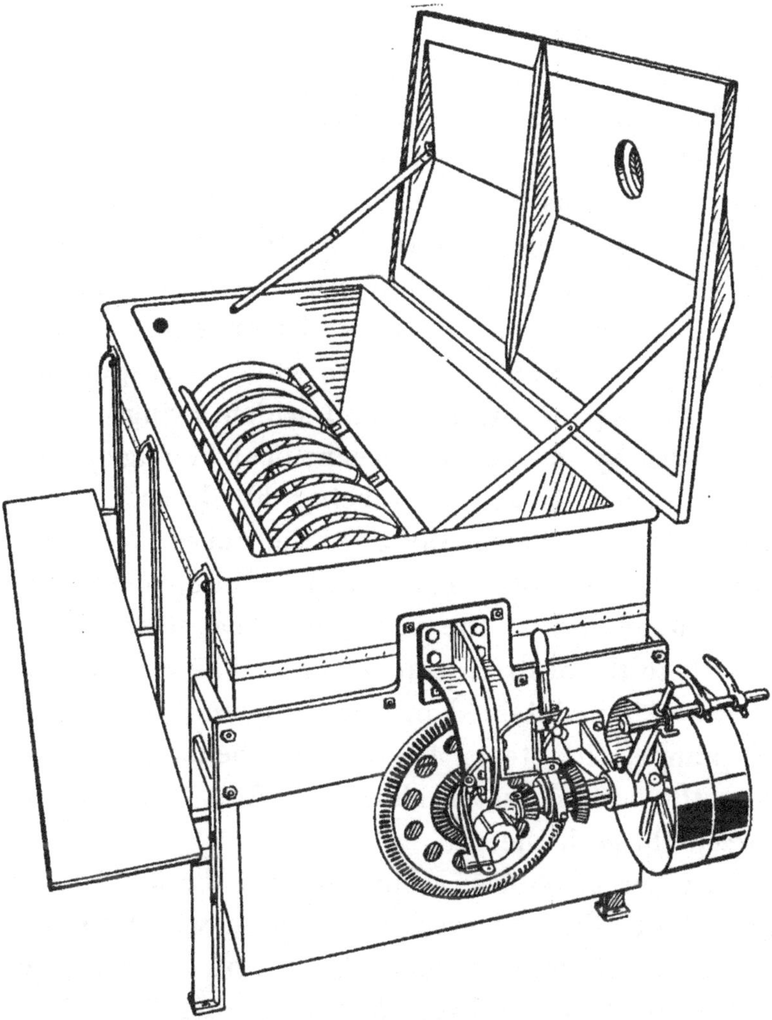

Fig. 14-B

to circulate, usually by means of a pump. These tanks have, extending through them in a horizontal direction, some form of agitator. This may be a tubular coil, as in Figs. 12, 13 and 14-B, or stirring paddles, as in Fig. 15, or it may be coil suspended from above, as in Fig. 16. Some of the older forms have, instead of a coil of pipe, a series of hollow screw-shaped discs. In some, as in Fig. 14, there is a tubular structure which is suspended from the top of the tank and which swings back and forth by means of a mechanical arrangement. In operation, these tanks being filled with milk, hot water is forced through the tubes or discs above referred to. These coils rotate, keep the milk mixed and, at the same time, heat it and thus raise it to the desired temperature. The circulation of water is maintained in some machines by means of pumps, the steam for heating being admitted into the return water shortly before it is again forced through the coil.

In other machines, as in Fig. 12, the admission of air into the coil creates a suction which forces the water through without the use of pumps. In the form shown in Fig. 15, the heating is aecomplished in a rather novel manner. In the space

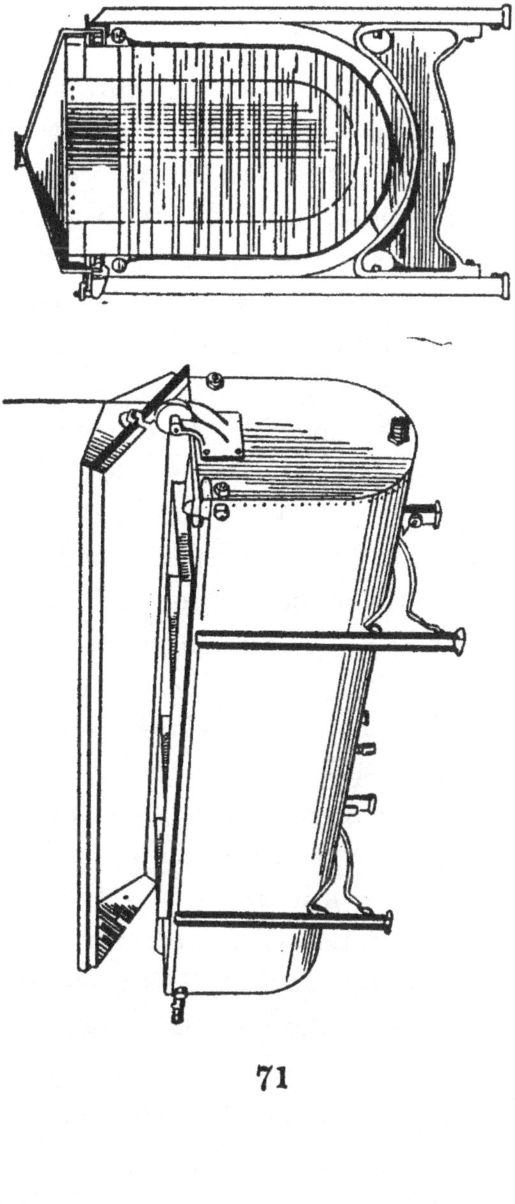

Fig. 15

between the inner and outer shell, there are placed near the top two perforated pipes, into which hot water is forced. The water is thrown through the perforations, in jets, against the outer surface of the inner shell. It flows down, following the contour of the tank, and drops from the bottom, from whence it is again pumped through the pipes after being reheated. The milk in the meantime is kept in agitation by means of the paddles.

In any of these tanks it usually requires from fifteen minutes to twenty-five minutes to raise the milk from 50° to a temperature of 145°. The proper temperature having been attained, the steam is shut off, and the milk is allowed to remain hot in the tank for any desired length of time.

The tank form of pasteurizer in practice shows excellent results. In plants where there are large quantities of milk to be pasteurized it is necessary that two, three or more tanks be arranged in series. This insures a continuous flow of milk through the cooling apparatus and the filling devices. In actual operation with three tanks, it occurs that while one tank is being emptied after heating and holding, the milk is being held in the second tank, while in the third tank the milk is flowing in and is being heated. These operations are repeated in rotation.

Many tests have shown that if the water in the heating coils is heated by injecting steam directly into the heating coil, the temperature at the inlet end is rarely above 165°. It may occasionally reach 175°. Usually it is sufficiently low so that the milk is not injured by superheating. It can be readily seen that the temperature at the inlet end of the coil is much greater than at the outlet end, since the cold milk takes up the heat, therefore the heating efficiency of the coil is not equally great for its entire length. Some attempts have been made to improve matters by placing a smaller perforated coil inside the larger pipe, so that the water distributed through the coil would have the same temperature in its entire length.

An objection to the submerged coil is that it becomes necessary to have the bearings packed with some sort of a stuffing-box. Stuffiing-boxes are always a source of possible danger both on account of the absorption of milk, which renders cleaning difficult, and on account of the fact that the constant attrition of the metal at the point of bearing creates a condition whereby a metallic taste may be imparted to the milk.

In the form shown in Fig. 15 the objections due to uneven heat distribution are at least partly overcome, since the heat is the same at both ends and on both sides of the tank.

In Figs. 15 and 16 there are no bearings in the milk, and stuffing-boxes are therefore eliminated.

It is obvious that when pumps of large capacity force the water rapidly through the coils, the temperature will be more nearly uniform throughout the entire length, and a lower initial temperature can be employed.

In plants where a comparatively small amount of milk is to be pasteurized, as frequently occurs in country shipping stations, one or two tanks will hold all the milk which is received at a plant each day. It is therefore possible to use the tank for several purposes. First, as a dumping tank for receiving the milk; second, as a heater; third, as a holding apparatus; fourth, as a cooler, since by replacing the hot water by ice water or brine in the coils, the milk can be cooled in the same tank. Fifth, cans can be filled directly from these tanks. In using the tank as a cooler, however, it is found

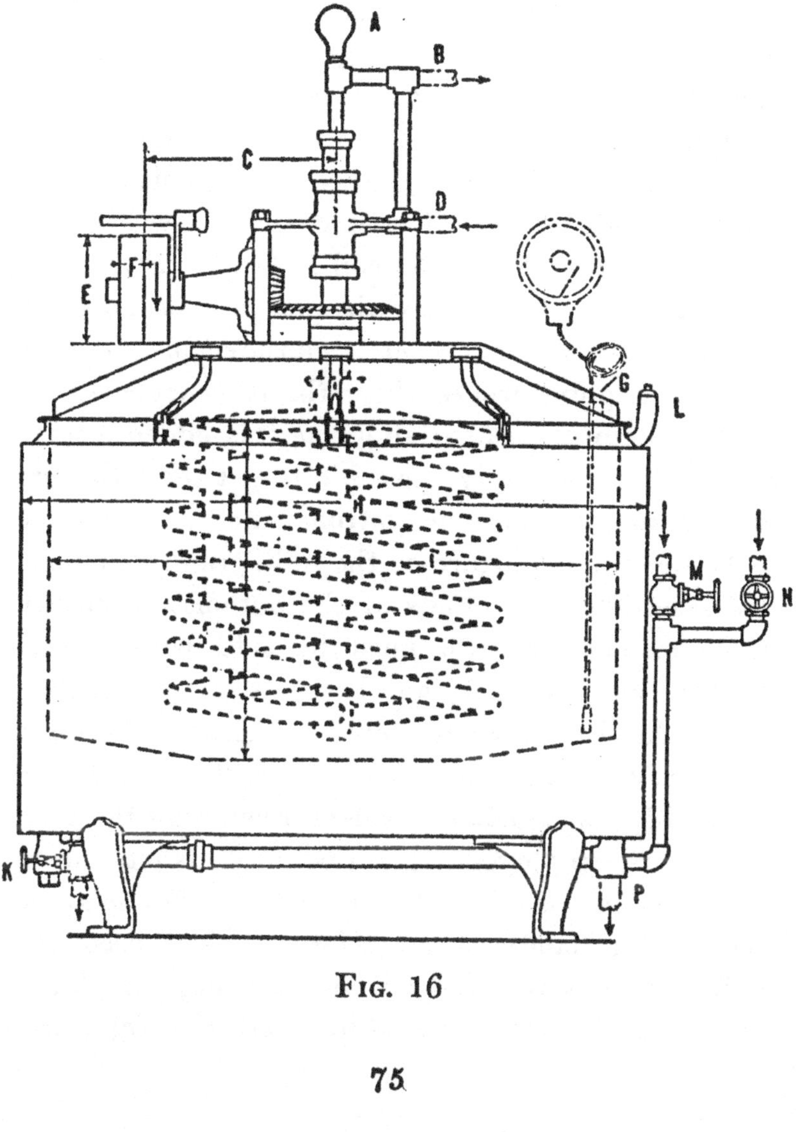

FIG. 16

that the time required is usually considerably over thirty minutes, which is longer than most authorities will allow. In some instances the writer has found that from an hour to an hour and a half was required. Where brine can be circulated rapidly through the coil, this time can be considerably reduced, especially if the coil is large. Some special tanks are provided with unusually large coils.

In plants which are at a considerable distance from the point where the milk is consumed, the tank system of pasteurizing has this added advantage. It is not always possible to get the milk into the creamery from the patrons and to pasteurize it before the departure of the milk train. In some localities the regulations of the authorities specify that pasteurized milk must be sold within from twenty-four to thirty-six hours after pasteurization. At these distant points it therefore becomes necessary to hold over some raw milk from day to day. If a pasteurizing system other than the tank system is employed, this raw milk is stored in cans, and a considerable amount of extra labor is required to handle them. If the tank is used, the milk which is received on one day may be placed in the tank and there stored till the following morning, it being kept cold in the meantime by the circulation of cold water through the coils. This milk may be heated on the morning of pasteurization as early as necessary to get it ready for shipment, no handling or transfer of the milk being previously necessary.

In case the tank is used as both a heater and a cooler, there is of course a considerable strain upon the joints of the apparatus due to the expansion and contraction of the metal. Those forms of apparatus are therefore the most desirable which are the most strongly constructed, and are therefore least liable to break. A coil of pipe is stronger than a screw disc.

Fifth Type.—Another process which has been used to a limited extent for pasteurized milk is known as bottle pasteurizing. By this method, the bottles are filled with cold milk and are hermetically sealed. They are then either placed in a tank, where they are submerged in water, or else they are carried under showers of water. In either case the temperature of the water increases as the bottles are carried forward. When the desired temperature is reached, the bottles are then held for the desired time and are then brought in contact with cold water, either in tanks or in showers, till the temperature is sufficiently reduced.

There is little doubt that from the purely theoretical point of view the pasteurization of milk in the bottle is the ideal method. During the process not only the bacteria contained in the milk are destroyed, but also whatever bacteria may have been contained in the bottles and on the caps are also killed. It can also be seen that the danger of recontamination after pasteurization is reduced to the minimum.

There are several forms of apparatus on the market in which the pasteurization takes place in the bottle, two of which are shown in Figs. 17 and 18. This method of treatment is very extensively used in the manufacture of beer. In Fig. 17, the cases filled with bottles of the milk are carried by an endless chain conveyor through several tanks of water in succession. The water in the first tank is warm, is hotter in the next, and so on till the maximum is reached, after which the milk is carried into the cooling tanks. The entire process requires an hour or more.

In the process shown in Fig. 18, an endless plat-

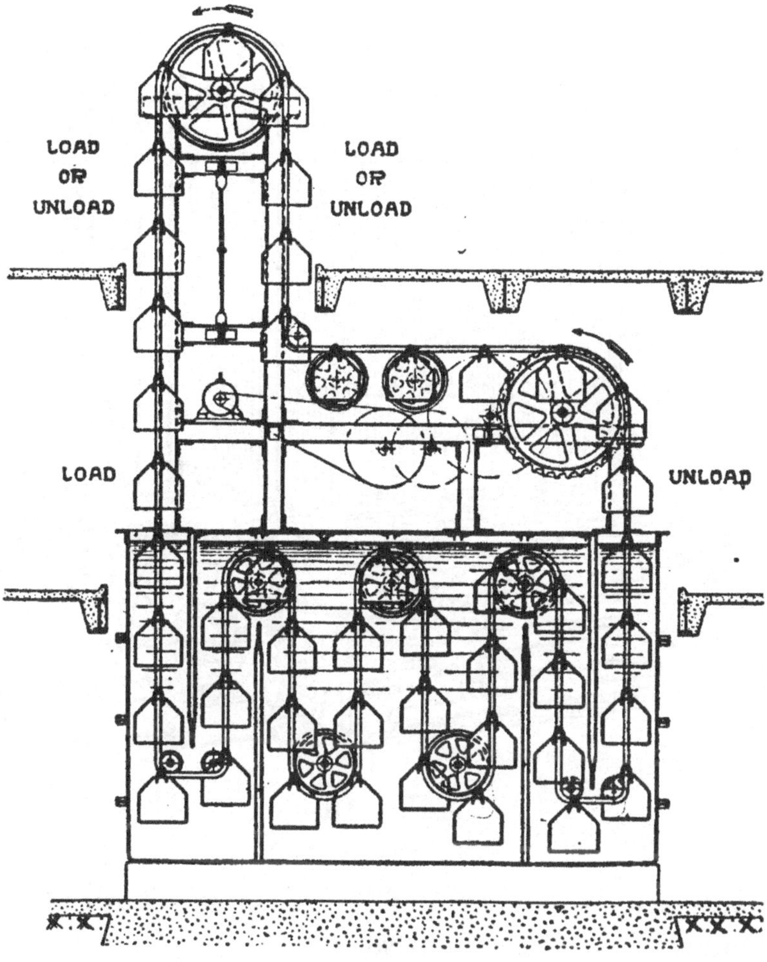

FIG. 17

FIG. 18

form carries the cases of bottles under showers of water, of varying temperatures.

When milk is pasteurized by this method, a more complete cooling is necessary than is the case with beer. More cooling tanks or additional showers are therefore required.

In a rather new machine, not shown in this book, the bottles are placed in a thoroughly insulated box or compartment, where they are subjected to a shower of water, the temperature of which is increased or decreased at will by proper regulating devices on the outside of the apparatus. It is claimed that with this outfit the heating and cooling is rapidly accomplished, and that the complete insulation of the compartment prevents the loss of heat. It is also said that, for the same reason, the same apparatus can be used as a cold storage box, and the treated milk can be allowed to remain in the apparatus after the completion of the process till it is ready to be taken out for delivery. These claims would seem to make it an especially desirable outfit for small dealers, and for dairymen in or near the cities where the milk is to be sold.

The objections which are advanced against the process of bottle pasteurization are, *first,* that the cap which it is necessary to use upon the bottles in order to secure a tight seal is expensive; too much so to warrant its use upon milk which is sold at the popular price. It can, however, be used on bottles containing milk sold at advanced prices, like Certified Milk, Grade A, etc. The *second* objection is that since the milk expands upon being heated, the bottles cannot be entirely filled. The bottles must therefore be made of a size especially large, and customers must be taught that bottles which are not full do still in fact contain the full measure of milk. A *third* objection advanced has been that milk in bottles which are not full is more likely to become shaken up, and the cream disturbed, than is the case in bottles completely filled. This objection is probably not of serious moment. A *fourth* and rather serious objection is that the expense in the consumption of heat units is much greater than when other methods of pasteurizing are employed. The reason for this is that the mass of glass in the bottle nearly equals the mass of the milk, and that since both bottle and milk must be heated and also cooled, nearly double the amount of heat is necessary, and a proportionately great amount of cooling medium is used. In large plants this added expense is very considerable.

An attempt has been made to overcome this last

objection, in part at least, by providing a series of tanks shown in Fig. 19. The five tanks shown in line No. 1, 2, 3, 4, 5 are used for containing the cases of filled bottles which are to be heated. These are lowered into the tanks by suitable devices. The other three tanks behind the pasteurizing tanks contain the water to be used in the process. One, A, contains hot water, which is kept at a constant temperature by means of a thermostat. From this tank the water is circulated by means of a pump through the tank, which at the time contains the bottles which are being heated and held.

A second tank, C, contains cold water, which is kept cold either by ice or by a brine coil. This water is circulated by a pump through the pasteurizing tank, in which at the time are contained the bottles which were being cooled.

The third tank, C, is called a tempering tank. The water in this is used for regenerative purposes, in the following manner: After a tank full of bottles has been heated and held for the desired length of time, the hot water is all returned to the heating tank, rendering the bottle tank empty of water. Cool water from the tempering tank is then admitted. This cools the hot bottles of milk to a certain extent, and is in turn partly heated. The water is then returned to the tempering tank, after which the cold water from tank C is pumped into the tank full of partially cooled bottles, and when the cooling is complete, the bottles are removed, and the tank is filled with fresh, unpasteurized bottled milk. The warm water from the tempering tank, then being admitted, it begins to warm the cold bottles and is itself cooled. After its return to the tempering tank it is ready to be used again in the routine which has been described, the process being repeated indefinitely. In this way a large amount of heat is saved which would be otherwise wasted.

From the illustration and the description the impression is gained that this apparatus would be rather cumbersome to operate, although it would no doubt be possible to somewhat simplify the mechanical arrangements for shifting the water from tank to tank. The writer is not personally familiar with any plant where such an arrangement is in use.

Another method has been devised for saving some of the heat otherwise wasted. This consists

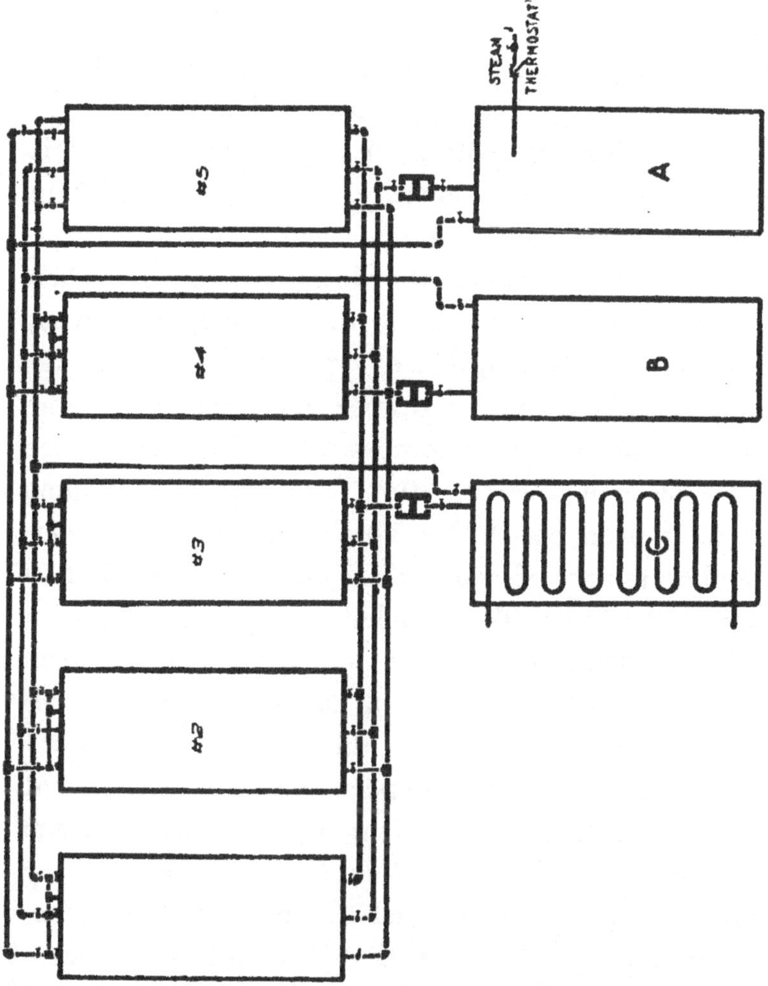

FIG. 19

in heating the milk in any good form of pasteurizing apparatus, and then pouring it, while hot, into bottles which have just emerged in a hot condition from this sterilizing apparatus. The heat which is usually lost in cooling the hot bottles is in this way saved. Since, however, the bottles must in any case be cooled, it can be seen that the cooling, which usually takes place slowly in the air when washed bottles are stored in the ordinary way, must now be accomplished rapidly in order to reduce the milk to the proper temperature in the time required. This necessitates the use of ice water or brine refrigeration, and since the volume of glass used equals the volume of the milk, double the amount of cooling medium is required than when only the milk is cooled.

Experiments were made by Ayers and Johnson in which the heated milk in bottles received a primary cooling by means of blasts of air at ordinary temperatures. These experiments indicated that a considerable economical saving of cooling medium thus could be effected. Completing the cooling by means of blasts of chilled air produced satisfactory results. The fact that the cooling process required a long time seemed to produce no harmful effects upon the milk. There was no noticeable increase in bacterial content when the cooling was accomplished within two hours. These results were obtained under conditions which somewhat approached what might be termed laboratory surroundings, and it is possible that under the stress of commercial conditions, difficulties of operation might develop which would unfavorably influence the results.

There is one objection which has been raised in the practice of pasteurization in the bottle which may have some weight in certain circumstances. It is claimed that because of the fact that the bottles are tightly sealed, there is no chance for the escape of disagreeable or offensive odors. In other heating and cooling processes, a certain amount of aeration takes place, with, as claimed by some dealers, beneficial results. Those who advocate bottle pasteurization claim that during the process of heating, the objectionable gases are driven out of the milk and are forced beneath the bottle cap into the outer air. In so far as known to the writer, the correctness of this claim has not been demonstrated. Neither is it known that objectionable odors or flavors have been subjects of complaint among the users of milk pasteurized in the bottle.

An objection which may be a serious one at times is this. If the bottle caps are not absolutely tight, as may occur when bottle necks are chipped or uneven, there is an opportunity for water to be drawn into the bottle. When the cooling is being carried on there is, of course, a partial vacuum produced in the space above the milk, and if in cooling the bottles are immersed in water, then the water might be drawn into the bottle. If the water were impure the danger becomes apparent.

A certain advantage in the use of the bottle pasteurizing process from a commercial point of view becomes apparent when only small amounts of milk are to be treated. If individual dairies are compelled by legal requirements to pasteurize the milk sold, it is a comparatively inexpensive matter to equip for pasteurizing in the bottle. The only equipment necessary is a tank large enough to immerse the filled bottles; facilities for supplying hot water to the tank, and also for replacing the hot water with cold when the heating is accomplished. Heat-recording devices are manufactured for use in connection with bottle pasteurizing. When this method is installed in small dairies there is saved the expense of installing heaters, holders, coolers, pumps and piping. Equally important to the dairyman is the fact that the operating expenses are less, since the amount of machinery to be kept cleaned and in repair is almost nothing.

A process of pasteurizing which, so far as known to the writer, has been but little employed, consists in making use of the vacuum pans which are commonly employed in the manufacture of condensed milk. With certain changes, these can be used satisfactorily in pasteurizing milk. As usually constructed, these vacuum pans are retort-shaped structures made of copper, and contain at the bottom two or three coils of copper pipe, into which live steam is admitted. When milk is condensed, these hot pipes raise the milk to the desired temperature, and the natural water contained in the milk is drawn off in the form of steam by means of the vacuum pipe attached to the top of the pan. If these pans are used without change as pasteurizing tanks, it is probable that the live steam in the heating coils would have a tendency to scorch the milk, and it would be necessary to replace the steam by hot water, which should be rapidly pumped through the coils, the water being maintained at a uniform temperature in tanks outside the condensing pan. The pans could be used also for holding the milk after it is heated for any desired length of time.

In plants where seasonal conditions require that the milk be used for different purposes at different times of the year, such an arrangement would effect a saving of expense in equipment, and would thus aid in obtaining greater operating efficiency.

From time to time attempts are made to bring forward some new process for treating milk designed to reduce the number of bacteria in milk by some agency other than heat. One of these is the treatment of milk by means of an electric current. In one electric process, the milk is allowed to flow from chamber to chamber through orifices of small diameter. While so flowing it is subjected to a strong alternating current of electricity. It is claimed that this current destroys the bacteria contained in the milk through the shock to which they are subjected, and that the digestive qualities of the milk are not impaired, but are rather improved.

It is, however, found necessary to heat the milk to about 130° before subjecting it to the electric current, and during electric treatment it attains a temperature of about 158°, due to the resistance of the milk to the electric discharge. It is an open question whether the destruction of bacterial life, which undoubtedly takes place, is due to the electric current or to the heat to which the milk is subjected. The results of tests made of the milk so treated indicate that the percentage of bacteria destroyed is high. If the effectiveness of the apparatus depends upon the heat generated during the process, it is probable the high percentage of bacterial destruction is partially due to the fact that milk flows in such a small stream through the machine. Every portion of it is thus uniformly heated to the maximum temperature, and even when milk is not held for any appreciable length of time, it has been found in laboratory tests that the bacteria are killed. As before stated, it is a fact that mechanical defects which exist in all machinery render it difficult to maintain uniform temperatures. This makes the holding of hot milk a necessary requirement in order to equalize temperatures.

The ultra-violet rays have been suggested as a means of killing bacteria in milk. Since, however, these rays have but little penetrating power, it has been found that the opaque character of milk prevents obtaining the good results which have resulted in the use of the ultra-violet rays in water purification.

The ozone process, which will by oxidation destroy some forms of bacterial life, and which will destroy odors, also acts upon the constituent parts of milk and so changes them that it seems probable that ozone can never be satisfactorily substituted for the usual pasteurization process.

An ingenious idea has been brought to the writer's attention by which it was proposed to destroy the bacteria by treatment in vacuum. It is well known that when persons work under abuormally high air pressure as in tunnel construction under rivers, etc., it is necessary for them to exercise the greatest care in coming out into the normal air pressure. The air, which under pressure is absorbed in the bodily tissues, suddenly expands when the abnormal pressure is relieved. The cells of the body may be strained or broken. In such case the disease known as "the bends" results. It has been proposed that a machine might be so constructed that an abnormal air pressure upon the milk could be suddenly released, thus allowing the air within the germ bodies to quickly expand and so destroy them. No such system has been made commercially available.

CHAPTER III

HOLDERS

As previously stated, the New York City Department of Health was the pioneer among health authorities to require that milk, in order to be officially considered to be pasteurized, must be held, after heating for a definite length of time. When this regulation was adopted no holding device was upon the market which was suitable to be used on a commercial basis. Since that time most of the manufacturing firms making milk-handling machinery have constructed and placed upon the market one or more forms of holding apparatus. These differ in detail, and to a considerable extent in the efficiency with which they operate. In general, however, it may be said they are naturally divided into two classes:

First, those which may be termed absolute holders, and

Second, those which are of the continuous or flow type.

Absolute Holders

The *absolute holders* are so called for the reason that the milk, after being heated, is all held in tanks or compartments for a definite length of time before being discharged.

The first holder placed upon the market was of this character, and consisted of an upright cylindrical tank having partitions radiating from the centre, thus dividing the tank into eight compartments. Such a holding apparatus is illustrated in Fig. 20. The drawing, however, shows a later type of the same apparatus, in which there are eight separate tanks which are arranged in a circle. An upright shaft extends through the centre of the tank or the group of tanks, as the case may be, to which is attached, at the upper end, a revolving spout, into which the milk is discharged through the pipe from the heater. This spout as it revolves causes each tank to be filled in turn. To each tank is attached at the bottom an outlet pipe and valve. These valves are kept closed by springs, and are opened by means of a revolving arm, or cam, attached to the lower end of the upright shaft before referred to. These outlet pipes discharge into a common pipe, through which the milk is conveyed to the cooler. By placing the revolving inlet spout in proper position with relation to the revolving cam at the bottom, it can be readily seen that each tank will be emptied after the milk has been held in the tank during the time required for the shaft to make a complete revolution. By properly gearing the mechanism which operates the shaft it can be seen that the time of holding can be fixed at will for any period desired. A little thought, however, will make it clear that the actual holding time will not equal the time required for the shaft to make a complete revolution, since approximately one-eighth of this time is occupied in filling each tank, and one-eighth is required to discharge each tank. In order to arrive at the actual minimum time for which we can be sure all the milk is held, we must reduce the time of revolution by one-fourth, or 25 per cent. Thus if it is desired to hold the milk for thirty minutes, it is necessary that the time of revolution of the operating shaft be forty minutes, which reduced by 25 per cent. equals thirty minutes, the actual holding time.

It is also necessary to make sure that the discharge pipes are completely closed before the milk is allowed to enter through the inlet pipe into any tank. Otherwise some milk may pass out of the holder without being held for any definite length of time.

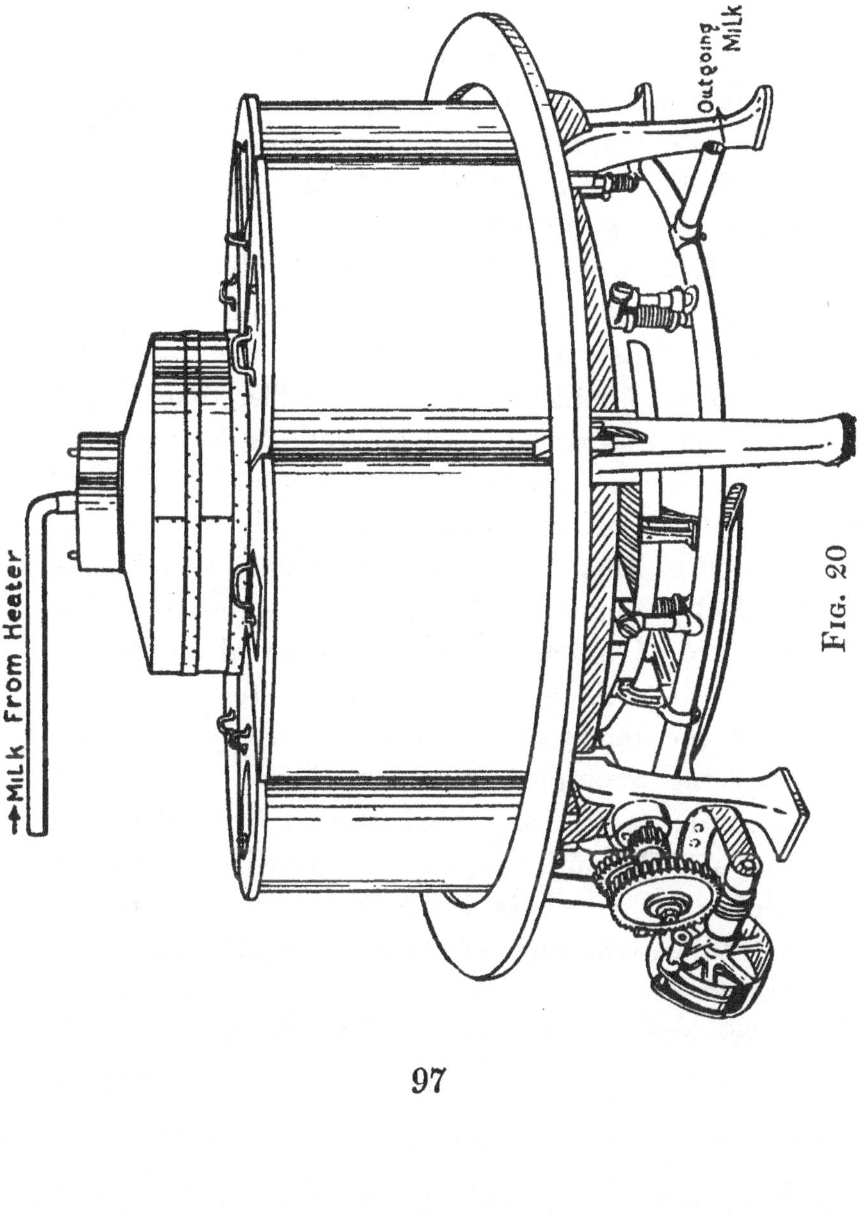

Fig. 20

Another point to be observed when partitions separate the tanks is to make sure that no tank becomes filled to overflowing. In such case, the milk will flow into a tank from which milk is being discharged, and thus the holding time becomes a matter of great uncertainty.

In this form of holder considerable foam develops, due to the dropping of the milk for a considerable distance from the discharge pipe into the body of milk in the tanks. The foamy milk does not retain the heat as well as is desirable, and as a consequence the destruction of the bacteria is not always sufficiently complete.

In Fig. 21 the eight compartments or wells are themselves immersed in a tank of water, the temperature of which is kept somewhat above that to which the milk is heated. In this form of apparatus the tendency to foam is largely overcome from the arrangement which allows the milk to enter the well from the bottom. Both the inlet pipe and the outlet pipe are operated by means of mechanism, which is above the tanks. There are sixteen valves, each of which is made tight by a ground seat, upon which the valve sets. These valves are at the lower end of the plug shanks,

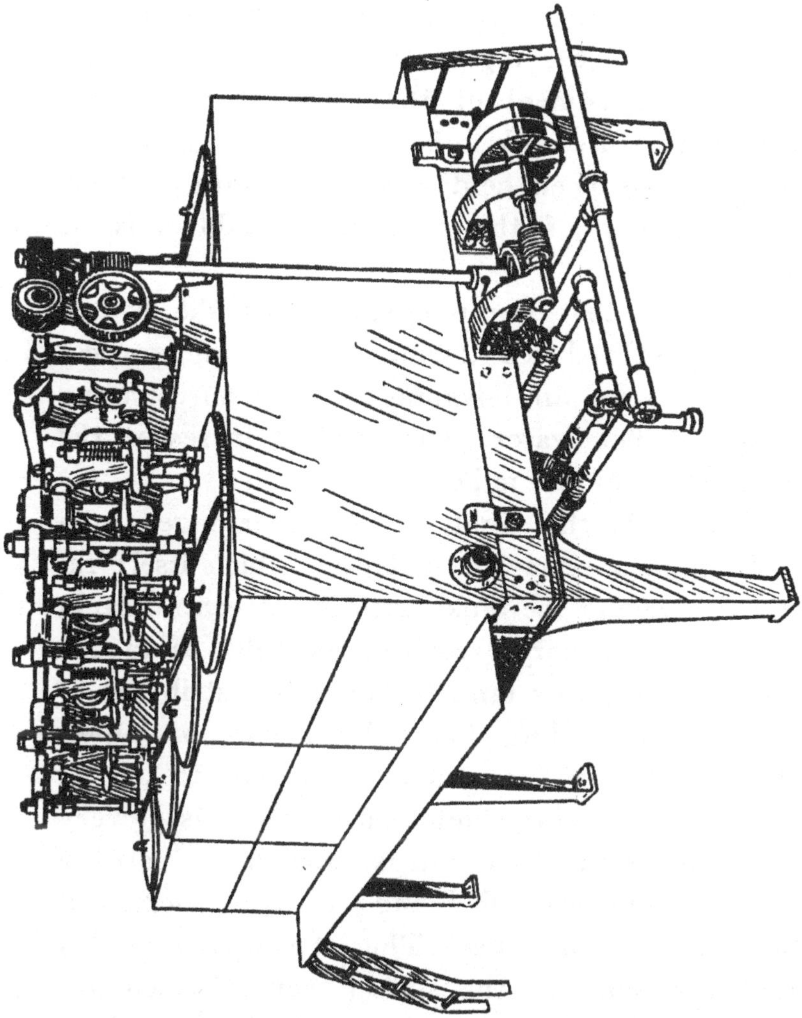

Fig. 21

which extend through the milk to the top of the tanks. Any loosening of these valves, or any inaccurate adjustment of the cams which operate them, may cause milk to be discharged from a tank before it has been held a proper length of time. The efficiency of the apparatus would thus be impaired.

The fact that the temperature of the heated milk is not reduced during the holding process appears to be an advantage. The surrounding jacket of hot water maintains or increases the temperature of the milk in the vats.

In Fig. 22 the eight tanks are arranged in a circle upon a frame or platform, which itself revolves, carrying the tanks with it. A central tank, which is stationary, receives the milk from the heater. From this tank the milk is distributed to the various holding tanks by means of pipes, which radiate from a central valve beneath the receiving tank. This single central valve is of very ingenious construction, and is so made that all the holding tanks are filled and are also discharged through the same valve. Thus it is seen that this one valve replaces the sixteen valves which are employed in such an apparatus as shown in Fig.

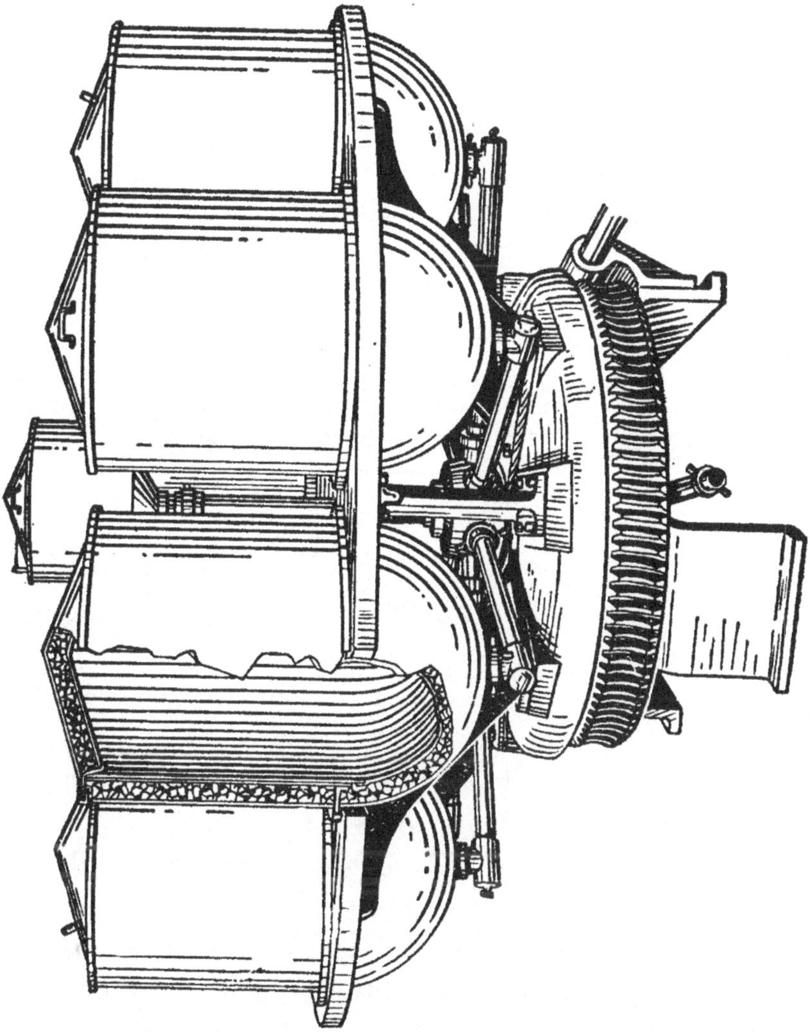

Fig. 22

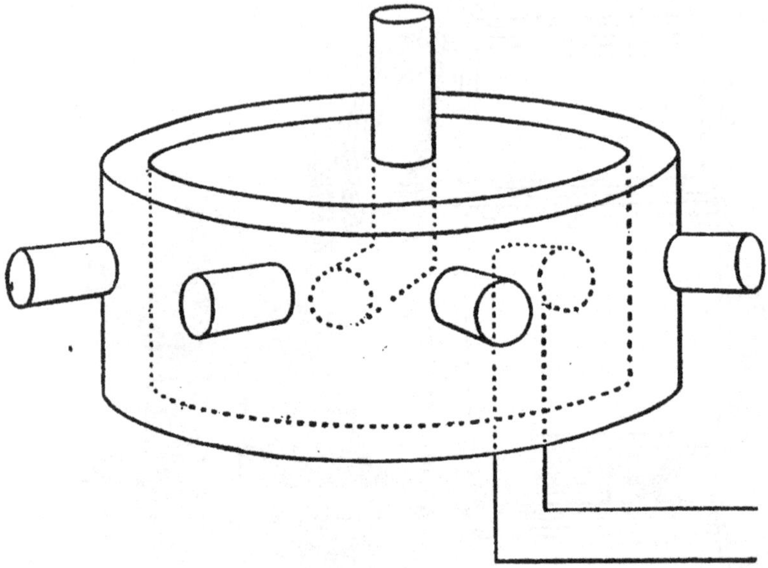

FIG. 22-B

21. The chances of valve difficulties are thus reduced.

The arrangement of this valve is rather difficult to describe. It is shown in Fig. 22-B. A central core is attached to the pipe, which descends into it from the central tank. This core remains stationary, and is provided with a single opening, which is a continuation of and is at right angles to the discharge pipe from the tank above it. Outside this core, and closely fitting it, is a shell which revolves around it. Into this shell the eight pipes from the various holding tanks are fastened, and they are so placed that as the shell revolves each pipe is in turn brought opposite to the opening into the inner core. The milk is thus admitted to the bottom of each holding tank in turn. As the shell continues to revolve each pipe in the outer shell passes in turn the inner opening and the milk supply is thus shut off from each particular tank, and so remains until the shell has nearly completed its revolution. The pipe opening from the holding tank then comes opposite to another opening in the inner core, which extends only a short distance into the core. This opening turns downward and is connected with a pipe, which discharges the milk to the cooler. Thus it is seen that each tank is filled in turn and is in turn discharged automatically. The same method must be employed for arriving at the minimum holding time as that used with the device illustrated in Fig. 20, namely, the time of revolution must be reduced by 25 per cent.

The exact holding time may be arrived at in another way, namely, by carefully noting the exact time at which the last milk enters any single compartment of the holder, and then observing the exact time at which the first milk flows from the same tank. The difference indicates the minimum holding time. Of course, some of the milk is held for a longer time than that which is thus indicated, but we are sure that none is held for a less time.

It will be found that the time as observed in this way is nearly equal to that which we obtain by calculation, allowances being made for variations in the mechanical exactness with which the machine operates.

It will be seen from the illustration that the various holding tanks are insulated by a cork jacket, so that little heat is lost during the holding period. It will be noted, however, that the pipes leading from the holding tanks to the central valve are not insulated, and in this condition the milk which lies in these pipes during the time of revolution may, and probably will, become cooled, and the effectiveness of the bacterial destruction will be impaired. This having been brought to the attention of the manufacturers, they provided insulating jackets for the pipes as well as for the tanks. Discharge pipes constructed with these insulating jackets are now provided for machines which are placed upon the market at the present time.

A modification of this type of holder is now being manufactured in which the holding tanks do not revolve but are made stationary and, instead, the central valve revolves. The same effect is secured as in the other type, and there is this advantage, that less power is required, and the heavy frame required to hold the revolving tanks is not necessary. There is also this added advantage, namely, that a small plant may install such an apparatus with but two or four tanks, and as business expands there may be added other tanks, it being only necessary to change the shell of the valve which revolves about the core.

In Fig. 23 there are seven oblong enamel-lined tanks which are immersed in a tank of hot water. These are each filled in turn by means of valves in a pipe which extends along the upper side of the series. These valves are opened and closed by means of an automatic arrangement operated by a revolving shaft extending parallel to the inlet pipe. A similar shaft extends along the discharge pipe at the bottom of the series, and operates valves which automatically regulate the discharge of the milk.

There is one thing which should be carefully observed regarding this form of holder. It will be noted that the two shafts which operate both the inlet valves and the outlet valves are impelled by a single worm-driven gear. Thus both shafts move in unison. Care must be taken to make sure that no inlet valve is opened into any tank till the outlet valve is completely closed, otherwise some milk will flow out without being held for the proper length of time.

In calculating the holding time with this apparatus, it must be noted that since there are but seven tanks in the series, there is required one-seventh of the entire period of sequence for each tank to fill and one-seventh for it to be emptied. Thus the entire time of sequence must be reduced by two-sevenths. A little thought will make it clear that as the number of tanks in a series decreases in number, the time of revolution or sequence must be increased. With eight tanks a period of revolution of forty minutes reduced by 25 per cent. would equal thirty minutes. If, how-

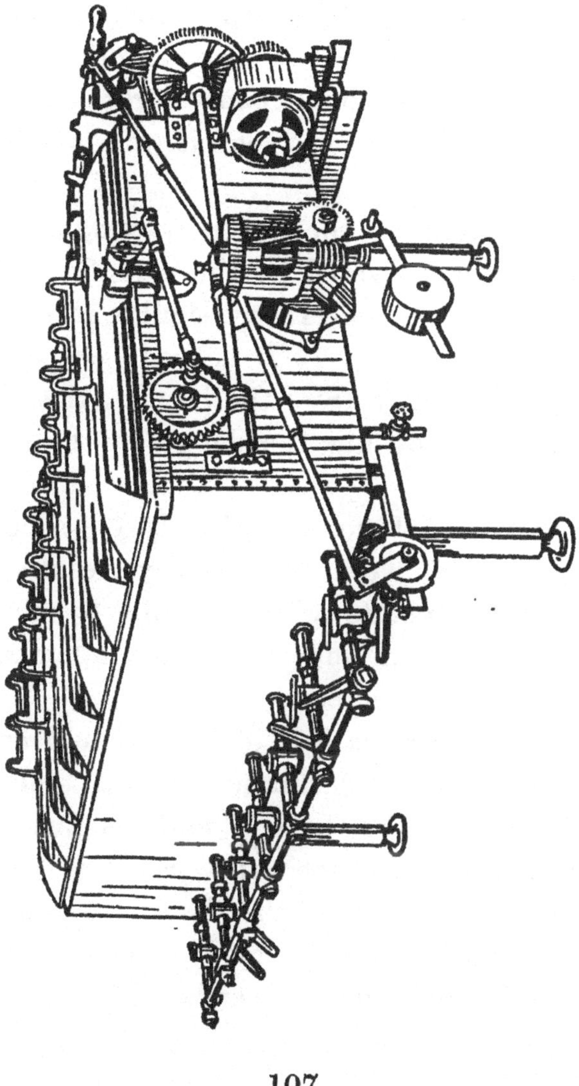

FIG. 23

ever, a seven-tank holder has a forty-minute sequence, which is reduced by two-sevenths, we have a holding time of but twenty-eight and four-sevenths minutes, and so on as the number of tanks decreases.

A modification of the valve system, shown in Fig. 23, is now provided by the manufacturers of this machine. A valve exactly similar in principle to that shown in Fig. 22-B is made use of. All tanks fill and empty from the bottom, thus avoiding foam and reduring the danger of by-passing the milk.

All the types of holders already described are being used with very satisfactory results. To summarize the points which should be observed in either buying or supervising the operation of holders of this character:

First.—Any serious loss of heat should be prevented by complete insulation of all tanks and pipes in which the heated milk is held. Of these insulations, the hot-water jacket is the most expensive and perhaps the best. There is a question whether the added expense secures enough added efficiency to render the expenditure wise.

Second.—We should be sure that every particle of milk is held for the minimum time required.

Third.—The valves for admitting and releasing the milk must be so tightly fitted that no milk can by any possibility be discharged before the proper holding time is completed. It may be possible for valves to become so worn by use that annular grooves will be worn around them, thus allowing milk to be by-passed while the valve is closed.

Fourth.—Make proper calculation to determine that the cycle of filling, holding and emptying of the tanks is long enough to make sure that the holding time is sufficient and in addition to allow for the filling and emptying of the tanks.

Fifth.—Make sure that the inlet valve is never open into a tank before the outlet pipe is closed.

Sixth.—Since the ease and convenience of cleaning is an item of importance, as a general propositiou it may be said that, other things being equal, a machine which has the fewest pipes and valves consistent with efficient operation is the most desirable.

Seventh.—It is desirable that as little foam as possible be produced in the operation of a holder.

Whatever form of tank holder is used, it is necessary that it be thoroughly sterilized before use each day, by the admission of live steam for a considerable period of time. This sterilization is of importance not alone for the destruction of the bacteria contained in it, but also in order that the pipes and tanks may become thoroughly heated immediately before use. If they are not so heated, it will be found that the milk first admitted to the apparatus will be considerably cooled, and will be discharged at a temperature much below that required by the regulations, and will so continue until the hot milk has itself slowly heated the apparatus. This may require fifteen or twenty minutes. It will be found under such circumstances that the milk first discharged from a holder contains many more bacteria than that which flows from it after the machine is heated.

Another form of holder which belongs to the absolute type is that which is represented by the tank heaters shown in Figs. 11, 12, 13, 14, 14-B, 15, 16, and previously described. In these the milk, after having been heated, is allowed to remain for any desired length of time, and is then released. This method of holding has produced very satisfactory results. It has this advantage; that the milk comes in contact with no new metal or apparatus between the completion of the heating process and the time when the milk is cooled. One disadvantage is that since it requires considerable time for a large tank full of milk to be discharged, the last milk leaving the tank is held for a period considerably longer than that which first flows from it, and this long holding tends to interfere with the proper rising of the cream.

Continuous Holders

The continuous or flow type holders differ from the absolute holders in that the milk, instead of being quietly held in retainers for a definite length of time until the required period for holding is completed, is simply retarded in its flow. This period of retardation is so extended that before the milk passes to the cooler it has retained its temperature for the length of time desired.

There are several forms of continuous holders. One of the first of these is the so-called Park holder, invented by Dr. William H. Park, of the New York City Department of Health, and pat-

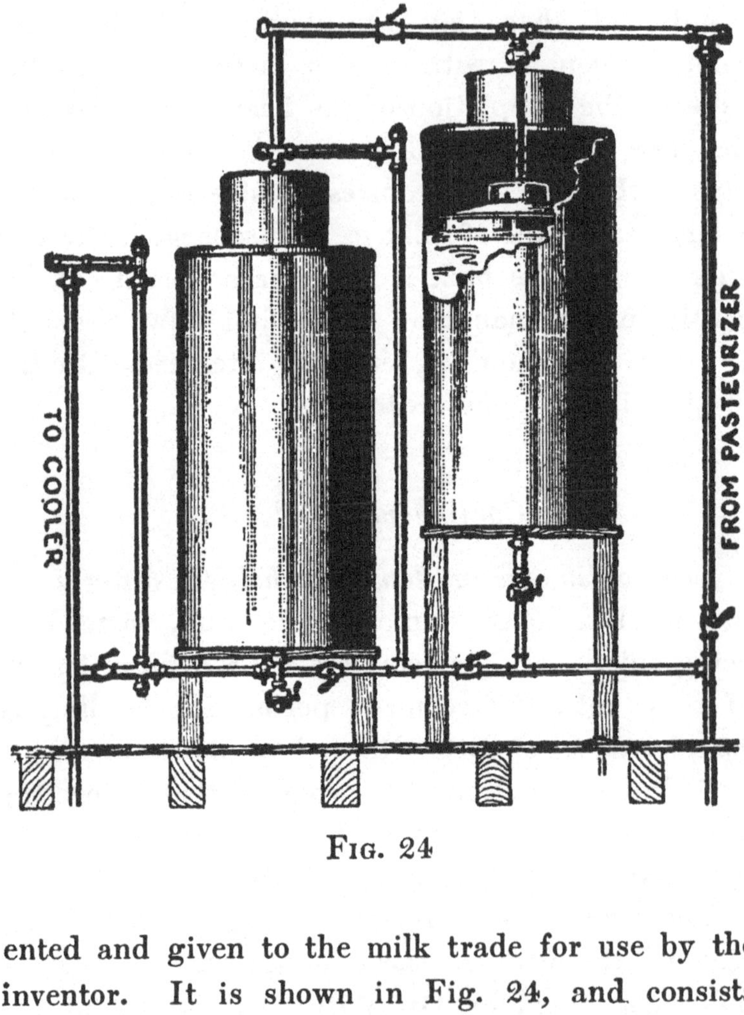

Fig. 24

ented and given to the milk trade for use by the inventor. It is shown in Fig. 24, and consists primarily of an upright cylindrical tank, into which the milk flows at the top, and from which it is discharged at the bottom. An inverted gooseneck attached to the outlet pipe so directs the outflow that no milk can be discharged until the tank has become filled. In theory, the hot milk which enters at the top becomes gradually cooled as it flows downward, and so the last milk to enter the tank will be the last milk to leave it. If this were entirely true, then the time required for filling a tank would be identical with the holding time. In such case the various particles of milk would pass through the tank like a flock of sheep, one after another. Unfortunately, there are several disturbing elements which enter into the problem, and the particles of milk, instead of resembling sheep, more nearly resemble a flock of chickens in their behavior. In the first place, the milk which is nearest to the outside circumference of the tank becomes cooled more than that which is in the centre, and as it cools it increases in weight, and naturally falls to the bottom and goes out. This tendency is partly overcome by providing insulating jackets. In some cases, as in Fig. 25, the jacket is a hot-water covering contained in the space between the inner and the outer shell.

Again, it can be seen that if the temperature of the milk as it enters the tank from the heater is not accurately controlled, it will frequently occur that the temperature will drop, and this cooler milk will fall through the body of milk toward the bottom. It is also true that the force of the milk as it enters the tank from the heater will, unless this force is broken, tend to carry the milk deeply into the body of that already in the tank, and so create currents which will interfere with the actual holding time. This force is broken by placing some sort of a baffle plate under the inlet pipe. This baffle is sometimes so arranged that it floats on the surface of the milk, and rises as the tank becomes filled. In other cases a flat piece of metal is placed near the top of the tank, and may or may not contain perforations which allow the milk to flow through as well as over the edge of the plate. It is immaterial what form of spreader is used so long as the currents are broken up. It will be seen by referring to the illustration that the inverted U-shaped pipe for the discharge of the milk is provided with a by-pass valve, so that the tank can be emptied when the operation is completed for the day. It is important that this

U-pipe be provided with a vent at the upper end, otherwise a syphon will be created and the milk will all be rapidly drained from the tank.

In plants where a large amount of milk is pasteurized, and there is therefore a rapid flow, it is necessary to install more than one of these tanks in series, and in some instances three and four tanks are so installed, each succeeding tank being a little lower than the one before it, and the valves so arranged that the milk-flow is from one to the other through the entire series. This also helps to equalize any uneven temperature which may occur as the milk is discharged into the first tank.

With all these precautions to insure a steady and even flow, it is never safe to assume that the holding time is correct unless a test is made. Only two feasable means of so testing the tanks have developed. One is the color test and the other the temperature test. Neither are absolutely accurate, but are sufficiently so for practical purposes.

In the color test, the tanks are filled with water, heated to the same temperature as is the case when milk is pasteurized. The time of filling having been observed, a solution of some intense color, preferably methylene blue or uranin, is allowed to flow into the tank with the heated water. The time is observed between the time of its introduction and the time when the first trace of it appears at the outlet. In order to make the test as accurate as possible, care should be taken that a complete solution is made of the color, otherwise small particles of solid color may fall through the liquid and deceive the observer. The solution of any color is of course slightly heavier than the liquid in which it is dissolved, and the tendency is for it to fall after its introduction. Therefore, the solution should be made as nearly as possible of the same specific gravity as the water in the tank. If it can be made slightly less in weight than the water the result of the test will be rather better. Even with the greatest care, it seldom happens that two tests of the same apparatus will bring identical results. Unless the speed of flow is accurately gauged, the results will differ.

The temperature test is sometimes used to determine the efficiency of these holders. When this method of testing is employed it is the practice to fill the holders with water in the same manner as with the color test. The temperature of the water as it leaves the heater is then suddenly raised 5° to 10°, and the time at which the rise takes place is noted. A careful watch is then kept at the outlet of the holder to determine when any rise in temperature occurs at this point. The elapsed time indicates the time of holding. The temperature, instead of being raised, may be lowered a given amount at the heater, and the corresponding drop in temperature at the outlet noted.

The inaccuracy of this method of testing is due to the fact that during the long holding period there is a natural tendency for the changed temperature to become diffused throughout the surrounding liquid. It can also be seen that when the temperature is raised the natural tendency is for the hotter liquid to remain on top of the tank, while if the temperature is dropped, the tendency is for the cooler liquid to settle toward the bottom of the tank. A graphic chart can be made showing the result of these tests very clearly. Using the upright gradations of the chart to indicate changes in temperature, and the horizontal gradations to indicate periods of time, the sudden rise at the heater will be clearly shown by an almost perpendicular linc. The line showing the rise of temperture at the outlet will be comparatively gradual in its rise and will not reach its highest point for a considerable time. It will also be found that the proportional rise in temperature at the outlet will not be as great as at the inlet of the holder. When the temperature of the inlet is dropped instead of being raised, it will be found that the change of temperature at the outlet occurs in a much shorter space of time.

In choosing between the color test and the temperature test, it can safely be said that for the Park holder the color test is the more accurate. When the tubular holder is used, which will be described later, it is probable that the two tests are nearly equal in accuracy.

When two or more tanks are employed in series, it is desirable that the piping be so arranged that the lowest tank can be filled with milk first when the process of pasteurizing is commenced for the day, and then when the lower tank is full the upper tanks should be filled and the process continued in this manner. The milk will then flow through all the tanks in succession from the highest to the lowest. The reason for filling the lower tank first is as follows: When each tank is first filled, the milk of course falls from the inlet pipe at the top to the bottom of the tank, and afterward onto the surface of the body of milk in the tank as it fills up to the level of the outlet pipe. In cousequence there is considerable agitation of the liquid and currents are created. This may cause the last milk which enters the tank to be the first to be discharged, and our holding time becomes a matter of uncertainty. If, however, the lower tank is the first one filled, we are then sure that all the milk in this tank is held for at least the time required for the other tanks to fill.

This method of filling may be effected in either of two ways: First, a pipe may extend along the top of the tanks in such a way that by opening the proper valves the milk may be directed into either tank at will. Second, the pipe may be arranged at the bottom of the series in such a way that either tank may be filled, if the proper valves are opened and closed. Both arrangements of pipes are shown in Fig. 24.

It has been found by tests that this form of holder never holds the milk for a period of time as long as that which is required for them to be filled. Depending upon the form of the tanks and the method of construction, the percentage of holding time to filling time is from 25 per cent. to 70 per cent. It is usual to estimate the time as 60 per cent. of the filling time. It is true that tanks which are tall and of comparatively narrow diameter are more efficient than are short and wide tanks, the reason probably being that with long, narrow tanks the movement of milk is comparatively rapid, and there is less diffusion of the milk than when the movement is slow, as in the wide, short tank. Tests have been made which indicate that when the rate of flow is twelve feet or over a minute, there is very little diffusion of the liquid.

In all tanks of this character it is desirable, and indeed almost a necessity, that some sort of an insulating jacket be provided in order that there be as little loss of temperature as possible during the holding process. The reason for this, as before stated, is that the cooling of the milk which is near the outer circumference of the tank causes downward currents. Thus the milk which falls to the bottom on account of these currents is carried out of the tank before it has been held for the proper length of time.

The insulation may be provided by means of

jackets of asbestos, cork or felt, or a surrounding tank of water may be provided, as in Fig. 25. By means of the insulation, if properly applied, the heat will be retained and the milk as it flows from the tank will have a temperature but little below that at which it enters. When the hot water insulation is used, as in Fig. 25, the initial temperature of the milk may become somewhat increased during the holding period. This additional heating, it can be seen, has a tendency to cause the milk which may enter the tank below the average temperature to become heated and rise toward the top of the tank, thus increasing the time for which it is held. As a matter of fact, the water-insulated tanks have shown a good percentage of holding time as compared with the filling time when the color test has been applied. The test of their efficiency as measured by the destruction of the bacteria contained in the milk is also good.

This Park type of holding tank has been somewhat modified by various manufacturers in this respect, namely, by introducing the milk at the bottom and allowing it to flow out at the top instead by the reverse method of flow. When the inlet pipe from the heater enters the tank at the centre of the bottom, and the milk is forced upward till the tank is filled, the color tests made have indicated that their efficiency percentage was small. Especially was this true when the outlet pipe was at one side and near the top of the tank. It can be seen that the force which is necessary to impel the milk in at the bottom against the weight of the superimposed milk in the tanks must of necessity create currents. This will be more noticeable if the milk is forced through the heater by means of a pump. The impulses will create a somewhat uneven flow and the resulting currents will be greater. The discharge of the milk from one side of the top also tends to cause currents and an uneven overflow.

One manufacturer has so constructed his tanks that the inlet pipe enters at the top through the cover, is carried down through the body of the milk in the tank to the bottom, and then discharges the milk into the tank by means of an enlarged perforated foot somewhat resembling a sprinkler head on a watering can. In the plants where this apparatus is used the milk flows by gravity and there are thus no pump impulses. The

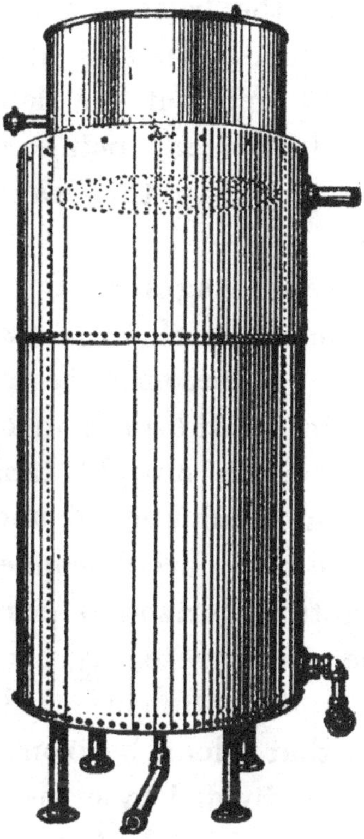

FIG. 25

force is broken up by the descent of the milk through the body of that in the tank. It can be seen that when the milk as it enters the tank is below the average temperature, the tendency will be for it to remain at the bottom, and so be heated before it is discharged. In this holder, also, the outflow is from the top, but the flowage into the discharge pipe is from many points at the surface into a gathering-pan. In this way surface currents are avoided.

One of the disadvantages of the upright tank holder is that when it is of large size it is rather difficult to clean. It becomes necessary for the workmen to get inside the tank in order to thoroughly clean it. Some upright tanks are being made which are mounted upon trunions in such a way that when the pipes are disconnected the tank may be turned into a horizontal position, and the cleaning can then be more easily performed.

A second form of continuous or flow type tank is so constructed that a long horizontal tank, which opens at the top, is divided by cross partitions into several compartments. At the top of each of these partitions is a gathering trough, extending along the upper edge of the partition. A tube attached to the centre of this trough extends to near the bottom of the next succeeding compartment. This is shown in Fig. 26.

In operation, the milk enters into the compart-

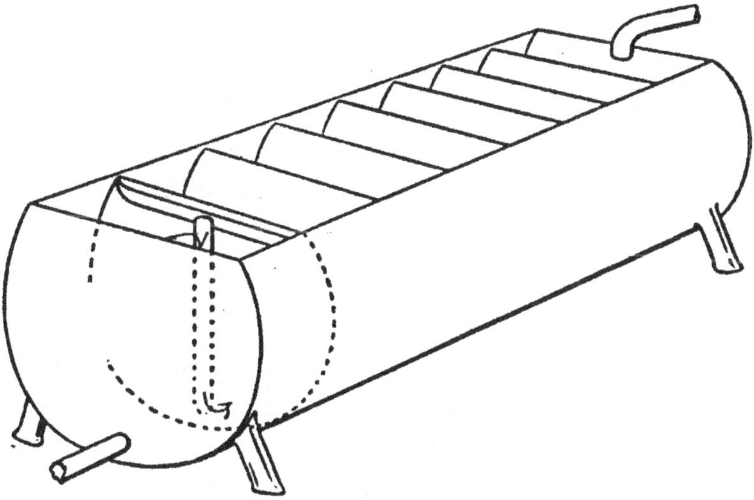

FIG. 26

ment at one end of the tank, and when the compartment is full, the milk flows into the gathering trough and is carried by means of the tube to the bottom of the succeeding compartment. Here the same routine is repeated till all the compartments are full. Then the flow becomes continuous till all the milk is pasteurized. When all the milk has left the heater, of course the flow from the holder stops. The tanks are at this time all full, and it is then necessary to remove the plug which is near the bottom of each compartment in turn till all are emptied. Since these plugs are removed by hand, it is necessary to place the hand and arm into the empty compartment at each operation. The danger of contamination is thus evident.

When the partitions are low it has been found that the milk flows over the tops of all of them, and there is thus but little holding time possible, since the milk flows almost directly from the inlet to the outlet pipe. Even with the best of construction, tests have shown that the percentage of holding time to the filling time is small, sometimes being as low as 10 per cent. to 15 per cent.

The third form of continuous flow type holder now used quite extensively is shown in Fig. 27. It consists of a series of large-size tubes arranged in a box-like structure. These tubes are connected one with another at the ends by suitable hinged heads, so that the milk which enters the top tube of the series flows forward and backward through all of them and is discharged from the lower tube. The discharge pipe is carried upward in the form of an inverted U tube in a similar manner to that shown in Fig. 24. This gives assurance that no

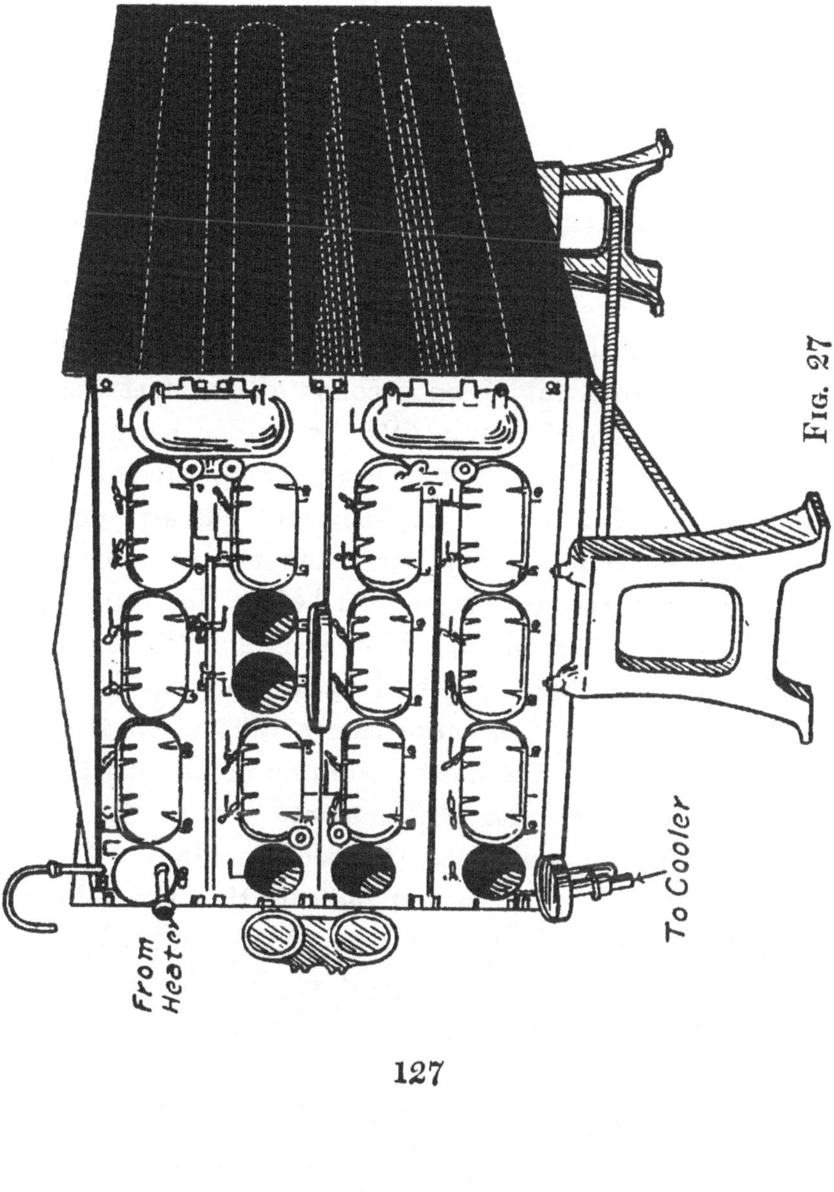

Fig. 27

milk will be discharged until all the tubes have become filled.

In this form of holder, the flow of milk through the apparatus is comparatively rapid. As before stated, this rapid flow tends to prevent diffusion, and the holding time more nearly approaches the time required to fill the apparatus than in any other form of flow-type holder. In some respects it is rather difficult to make a satisfactory color test upon this holder on account of the fact that the air, which of course fills the holder when the filling with milk is commenced, is not all forced out as completely as is desired, and thus air pockets are formed. It thus occurs that some milk is forced from the discharge tube before the apparatus is completely filled. Thus it is difficult to obtain the exact time required to fill the tubes. It is claimed by the manufacturers that when it is operating under the most favorable conditions the milk will be held for 98 per cent. of the time required for the apparatus to become filled. The tests observed by the writer indicate that the holding time is from 80 per cent. to 90 per cent. of the filling time, which is a greater percentage than with other holders.

The same difficulty which is experienced in emptying the tubular heater is observed with relation to the tubular holder. This is due to the very slight inclination or pitch of the tubes from the horizontal. In some cases it has been the practice to force the last milk out of the tubes by following it with water. The possibility of adulterating the milk unless great care is exercised is apparent. In the later machines made, the tubes are given a greater pitch, and the objection due to slow drainage is at least reduced.

It should be noted that unless the tubes are very thoroughly heated by means of hot water or steam immediately before use, the first milk entering the holder will be so cooled that its temperature at the discharge will be below that which is required.

This form of holder can be easily cleaned, since the tubes are large, and can be readily opened. It is thus easy to force a suitable brush through the pipes, and it is not difficult to see if all dirt has been removed. It is of course necessary that the cleaning be performed with great thoroughness in order that it be effective in the destruction of bacteria.

Whatever form of holder is used, whether it be

of the absolute or the continuous type, careful sterilization should be practised, not only directly after the apparatus is washed each day, but also immediately before its use on the following day. The reason for this is due to the fact that any organic matter which may remain in the apparatus after washing may contain spores of bacteria which are not readily killed by the heating to which the apparatus is usually subjected. These spores may develop into active bacteria during the time that the apparatus remains idle. Especially would this be true if, as usually occurs, considcrable moisture is allowed to remain in the machine. For this reason the holder should be thoroughly sterilized before use each day. It may here be noted that there is even a greater necessity for sterilizing the cooler, pipes, bottling machine, etc., with which the pasteurized milk comes in contact after leaving the holder. Any recontamination at these points is not afterward corrected and may be a source of danger.

It should here be noted that a feature which has been introduced into some milk plants consists in a brief superheating of the milk after it leaves the holder. This is accomplished by allowing the milk to flow in a thin layer over pipes which are heated to about 162° to 165°. The milk is then immediately cooled. It is claimed that bacteria which are resistant to 142° temperature are nevertheless considerably weakened by this heating, and are completely destroyed at the brief heating of 162°. It is also claimed that the cream rises in a more satisfactory manner when the additional heating is employed. The feature of applying a brief superheating temperature to milk already pasteurized has been covered by a patent.

It must be evident, when a little thought is given to the matter, that in order for any continuous holder to be reliable, the rate of milk flow must be uniform. In very few forms of apparatus is there any attempt to properly control the rate of flow. When milk flows by gravity, the rate at which it flows into the heater will depend partly upon the volume of the fluid in the tank which feeds the machine. When a pump is used to supply the heater, its rate of delivery may be influenced by the steam pressure, or by the degree to which the steam valve is opened, or by the condition of the valves and pistons. In places where a gravity flow is employed, it is not a difficult matter to arrange a feed cup which will make it certain that no more than a definite amount of milk can reach the heater per hour. When a pump is used, a float valve in the tank which feeds the heater can be made to shut off the intake pipe to the feed tank. This increases the milk pressure upon the pump and the discharge pipe, and by means of a suitable valve this increased pressure is made to close the steam valve which supplies the pump. This method of controlling the flow of milk to the heater is in successful operation in several pasteurizing plants.

CHAPTER IV

TEMPERATURE CONTROLLERS AND RECORDERS

In order that uniformly good results in the pasteurization of milk may be obtained, it is absolutely necessary that the temperature to which the milk is heated be controlled in such a way that there will be no great variation. In order to control this properly, three things are necessary—first, a uniform, even flow of milk; second, a steam pressure which does not vary; and third, that the steam which is admitted to the heater be so controlled that whatever variations there may be in the temperature of the inflowing cold milk may be met by varying amounts of steam, admitted to the heater. This control will, of course, also prevent any excessive heating which would follow if too much steam entered the heating apparatus.

The matter of an even milk flow has been discussed in the previous chapter.

The control of steam pressure may be accomplished in various ways. It is, of course, necessary that there be sufficient boiler capacity to insure that the steam pressure does not fall below a certain minimum amount. One of the simplest methods of steam pressure control is by an instrument constructed like the safety valve of a boiler. A movable weight upon an arm regulates the pressure which is desired. When properly set any increase of pressure above the desired amount lifts the arm and shuts the steam valve connecting with the pipe leading to the heater. Other more complicated forms of apparatus are on the market, the details of which it is not necessary to describe.

In the matter of the control of the temperature at which the milk leaves the heater, it is, of course, possible to do this by hand. This method makes it necessary that a man constantly watch the thermometer attached to the outlet milk pipe, and at the same time open or close the steam valve to adjust the steam flow to the varying temperature of the milk.

If the speed of milk flow, and also the steam pressure is well controlled, it is possible for a careful workman to regulate the milk temperature with good success. If, however, these factors are constantly varying, it is nearly impossible for even a careful man to obtain good results. It is, of course, more economically efficient to arrange for automatic temperature control than to depend upon man power, provided, of course, that the machine works as it is intended to work.

The machines designed to control temperatures are operated in two ways. In one machine the bulb which is inserted into the milk is filled with some volatile liquid like ether and is connected by means of a capillary tube with a pressure coil, which in turn regulates the flow of compressed air to a valve. The compressed air is employed as a motive power to operate the valve which regulates the admission of steam to the milk heater. Such a system involves the installation of an air pump and a tank for the storage of compressed air. To one unfamiliar with the apparatus the description may appear involved and the operation complicated. Fig. 28 shows this system.

In another form of controller, electricity is employed as a motive power in operating the steam valve instead of compressed air. The bulb which enters the milk contains two metals, which are so fastened together that the difference in expansion makes and breaks the electrical connection and the steam valve is operated by this means. This is shown in Fig. 29. This method can, of

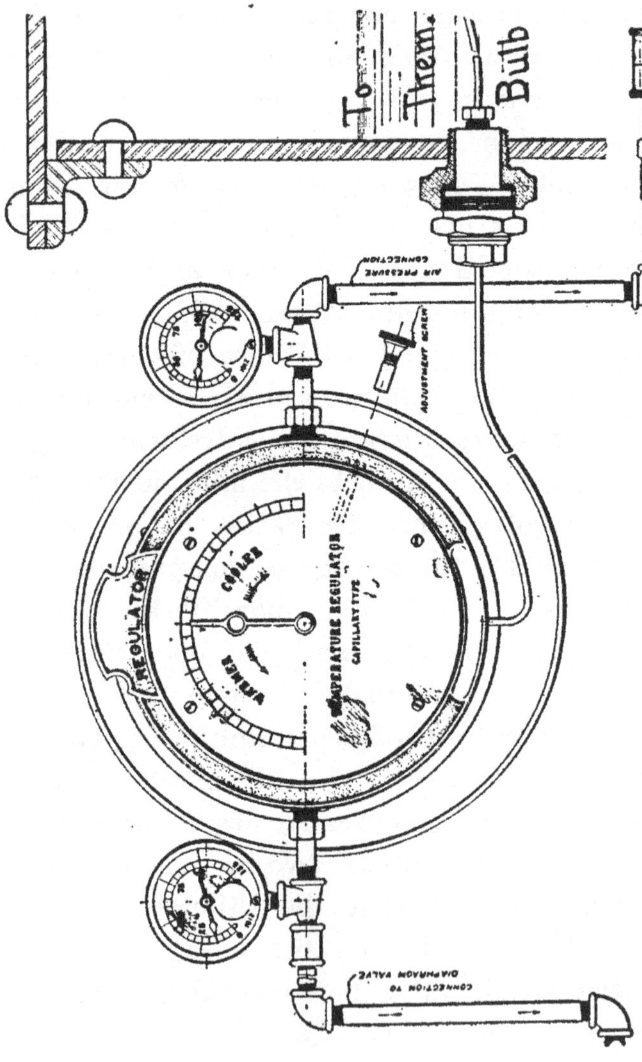

REGULATOR (CAPILLARY TYPE)

FIG. 28

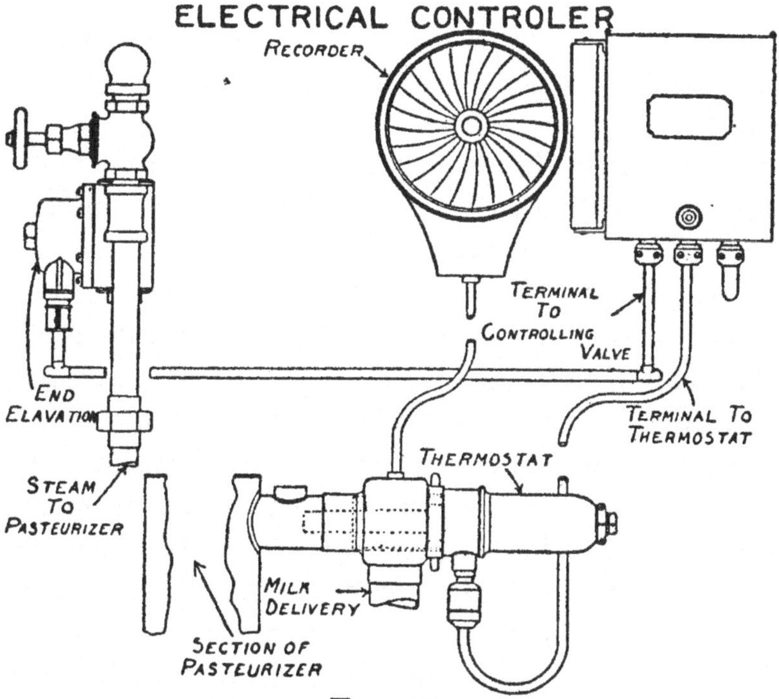

FIG. 29

course, be employed only where an electric current is available. This current, however, may be furnished by a small inexpensive dynamo, which can be operated at any plant where power is available. Both these forms of control work in a satisfactory manner. Of course, they are somewhat delicate in adjustment, and careful handling is necessary. Either form will control the milk temperature within 2°, as indicated by the temperature recorder.

A method of temperature control which was proposed by the writer some years ago had for its basis a principle which has been made use of in this connection to a limited extent only. This consisted in this, namely, that the temperature of the heating medium be so controlled as to remain constant, and that the milk which flowed out of the heater be kept at a uniform temperature by varying the speed of flow of the milk through the heater. This reverses or inverts the present practice, in which the speed of milk flow is constant, while the temperature of the heating medium varies. The proposed method would have this advantage, that no part of the milk could ever be heated higher than the temperature of the water, and if this water temperature were controlled by a thermostat no scorching of the milk would ever result. It is evident that with a uniform heating medium the milk which flows through the heater will take up more of this heat if the speed of flow is slow than it will if the milk moves rapidly over the heating surface. A thermostat attached to the milk outlet pipe from the heater could be made to control the steam which supplied the milk pump. This would vary the speed of milk flow. If milk flows by gravity to the heater, the thermostat might be made to open or close a valve in the feed pipe.

The nearest approach to the use of this principle occurs in some plants when the heated water is kept at a uniform temperature by a thermostat, and the milk flow is also constant. The heating surface of the heater is so very large compared to the amount of milk heated, that the water is never more than 3° or 4° hotter than the milk, and there is therefore no possibility of scorching the milk. In these plants also the efficiency of the regenerative principle is carried as near the theoretical limit of perfection as it seems possible. The water leaving the heater is within a few degrees of the temperature of the cold milk entering the machine. This cooled water is conveyed to the milk cooler, where it is used for cooling the hot milk. It becomes so heated when leaving the cooler that it is but a few degrees lower than the hot milk. It can be seen that but little steam is required to again raise the temperature to the desired point. The increased amount of heating and cooling surfaces required for this form of pasteurizing machine calls for careful cleaning and sterilization.

Automatic Recorders

The automatic recording of the milk temperature is desirable, and in most cities where the pasteurization of milk is subject to public control it is required. The milk company which is really anxious to put out a safe product wants an office record which will indicate whether the work is being properly performed. With health authorities, such a record is necessary, since it is impossible for inspectors to be at pasteurizing plants all of the time to watch their operations. A correct automatically made record is the best substitute for personal inspection. To be of ideal value, such a record should show, not only the temperature, but also the time of holding. When the tank or batch heater is used, the single recorder attached to the tank will show these items with fair accuracy, but the observer must note whether or not the cooling is either wholly or partially done in the tank itself.

If such cooling in the tank does take place, then the length of time for which the recording mark remains at the highest point is a correct indication of the time for which the hot milk is held. If the cooling is performed entirely outside the holding tank, then allowance must be made for the time necessary for the milk as drawn from the tank to fall below the point where the thermometer bulb is inserted into the milk. It is plain that the making of the record by the recorder pen will continue as long as the bulb remains in the hot milk. Thus, if the highest temperature were recorded for forty-five minutes, and it required twenty-five minutes for the milk in its discharge to fall below the recorder bulb, then the actual holding time for all of the milk would be but twenty minutes.

When the continuous holding process is used, it is desirable that a recorder be attached to the outlet of the heater, and that another be placed in the outlet pipe from the holder. If the charts on these two recorders are both set at the correct hour of the day then it is easy to calculate the duration between the time of the first discharge of hot milk from the heater and the first discharge from the holder. If one recorder only is in use, it is more important that it be placed at the holder outlet. It is evident that if there is a serious loss of heat during the holding process due either to imperfect insulation or to insufficient preliminary heating of the holder, then the milk would not be satisfactorily pasteurized, since some of the dangerous organisms originally in the milk may escape destruction. The holder recorder indicates this.

The various recording instruments upon the market differ in this one particular, namely, that in one form ether, or a similar volatile fluid, is used in the thermometer bulb and capillary tube, while in the other, the bulb and tube are filled with mercury. One of the former is shown in Fig. 30, while the mercury bulb is shown in Fig. 31. There is a variation also in the method employed for regulating the recording arm. In a form of recorder not shown, another form of adjusting screw is attached to the recording arm, and an improved device is used for securing the paper chart to the face of the instrument.

In the selection of a recording instrument, one of the most important points to consider is its sensitiveness to rapid or frequent changes in tempera-

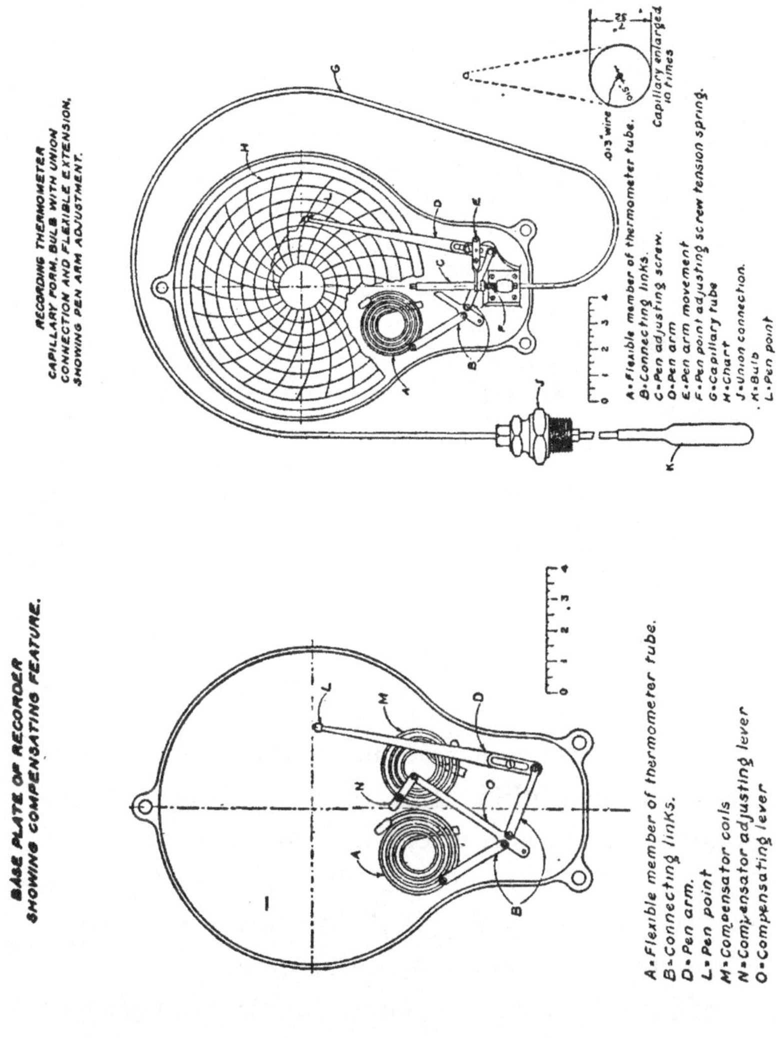
RECORDING THERMOMETER
CAPILLARY FORM, BULB WITH UNION
CONNECTION AND FLEXIBLE EXTENSION.
SHOWING PEN ARM ADJUSTMENT.
.013" wire
Capillary enlarged 10 times
A-Flexible member of thermometer tube.
B-Connecting links.
C-Pen adjusting screw.
D-Pen arm
E-Pen arm movement
F-Pen point adjusting screw tension spring.
G-Capillary tube
H-Chart
J-Union connection.
K-Bulb
L-Pen point
BASE PLATE OF RECORDER
SHOWING COMPENSATING FEATURE.
A-Flexible member of thermometer tube.
B-Connecting links.
D-Pen arm.
L-Pen point
M-Compensator coils
N-Compensator adjusting lever
O-Compensating lever

FIG. 30

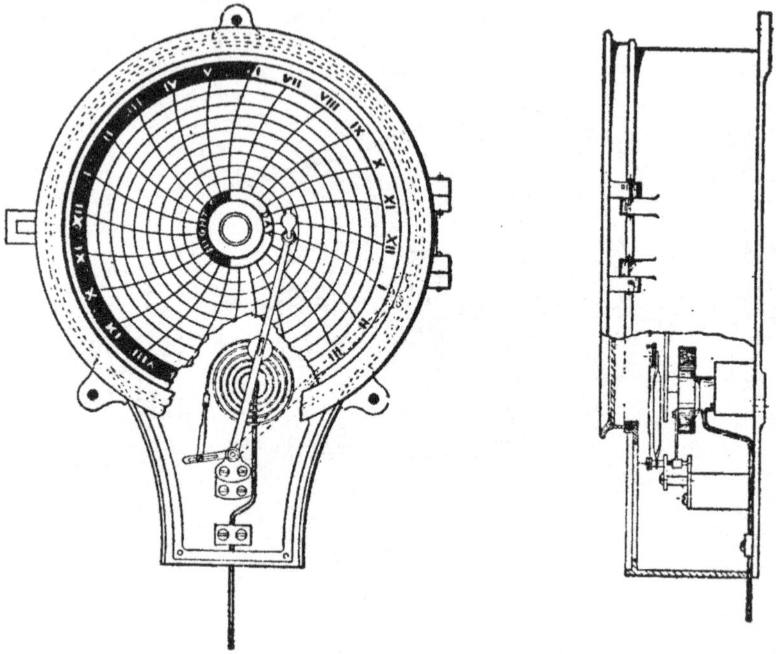

FIG. 31

ture. If it is slow in responding to such changes, there is a possibility that there may be a wide fluctuation of temperature, which will not be shown upon the recording chart. The manufacturer of each instrument claims that his machine is very sensitive to rapid temperature changes.

It should be borne in mind that a temperature

recorder is in reality a *pressure* recorder only in which the changes in pressure due to varying temperatures are so graduated as to indicate with fair accuracy the temperature of the milk. With the ether instrument, the amount of movement of the arm for, say, a ten degree rise, changes as the higher temperatures are reached. So the gradnations on the chart are not of the same width all the way across. In the mercury instruments this is not true, but the degree graduation lines are equally distant one from the other. There are certain inaccuracies in these instruments which must be taken into account. Thus it can be seen that the temperature of the room in which the recorder is placed must have an effect upon the pressure coil. In the instrument shown in Fig. 30 an attempt is made to correct this by attaching to the arm a compensating coil. This coil also contains mercury, and is so placed that it moves in a direction opposite to that of the coil to which the capillary tube is connected. The recording arm is attached to both coils. When, therefore, both coils move as the result of the room temperature, no effect is produced upon the recording arm, but it remains stationary. Any increased pressure in one due to the temperature of the heated milk will move the pen and cause it to make a proper record. The same effect may be produced in another way. In this case the compensating coil is constructed of two different metals, and the differences in expansion cause it to move, but in a direction opposite to that produced by the ether coil.

It must also be borne in mind that the recorder does not register absolute temperatures, but only relative changes in temperature. Thus inaccurate adjustment of the arm may produce a record which is deceptive. For this reason it is necessary to provide some means of adjusting the arm in order to secure an accurate reading. In some instruments the arm has a flexible joint which can be moved by hand. This is rather difficult to accomplish, without danger of breaking the recording arm, and fine adjustment is not easy. There is also the danger that the joint will be unintentionally moved when placing the chart upon the face of the dial.

In one recorder a post is so placed that it can be turned by means of a clock key. After the proper adjustment has been made, the post may be held in proper position by sealing it to other rigid posts beside it. This form is of value to the inspector, since by placing his seal upon it he can be sure that no change is made in the recorder during his absence.

In other instruments other means of adjustment are provided. If, as in some forms, the coil is so placed that its axis of rotation is the same as that of the recording arm, then no movement of the arm at the joint can affect the relative temperature valves of the varying degree spaces upon the chart. With this form there appears also to be less liability to vibrate with the vibration of the walls of the buiding in which it is placed. This form also has the advantage that the knob or screw which holds the chart to the face of the dial is so arranged that the chart cannot be left loose upon the apparatus, but must either be securely fastened or left off the instrument altogether. It is also very easy to fasten the chart to the clock face.

It has frequently been found by inspectors that instruments were not accurately adjusted. This may be due to a defect in the instrument itself, or it may be due to an intentional false adjustment by the man who is operating the pasteurizer. He may wish to indicate upon the chart a higher temperature than is actually employed in the pasteurization of the milk. For this reason those instruments whose adjustment is under the control of the inspector alone are desirable. They should be frequently tested. This frequent checking is of advantage to the honest milk dealer as well, since if a recorder by any means shows a temperature lower than it should, there may be trouble with the cream line due to an unsuspected high temperature in the pasteurizer.

Health authorities must be alert in watching the temperature records which are kept on file by the dealer. It is, of course, possible for a dishonest dealer to make false records. These may be made in various ways. Instances have been known where the entire record has been made by hand, an ordinary pen being used. This requires more skill than the average dealer possesses, if he is to deceive the wideawake inspector. It is possible to put a chart upon a recorder dial and leave it so loosely attached that it does not revolve with the clock. Then, when the desired temperature is reached, the chart may be turned by hand and a beautiful even line be drawn upon it. When the first milk from a holder has a temperature lower than required by law, some dealers wait until the proper temperature is attained, and then, by placing the thermometer bulb in the milk, they get a record which indicates an immediate rise of temperature to the highest point, and then a uniformly even line. Any record which indicates that the temperature has been raised at once to the maximum point, and then continues with no variation during the entire run, should be looked upon with suspicion. Inspectors with a little experience soon learn to detect false records, by evidences not always easy to explain, and they are able to gain from dealers the admission of irregularities which are at first denied.

CHAPTER V

CLEANING AND COOLING THE MILK—CLEANING CONTAINERS—HOME PASTEURIZATION, ETC.

It is of course needless to say that all milk should be clean when it arrives at the plant where it is to be pasteurized. That is, there should be no visible dirt in the milk. It is also desirable that all dirt which is not readily detected be absent. While visible dirt and bacteria are not identical, still it is true that visible dirt is almost always accompanied by bacteria in large numbers. Dealers should therefore make every effort to induce the dairymen to be so cleanly in their methods that very little dirt will find entrance into the milk. The sediment test when properly used is an excellent method of finding out whether or not the dairyman is cleanly in his methods. If, however, the producer is allowed to strain his milk before bringing it to the shipping station, then the sediment test simply tells the dealer whether or not the farmer has strained the dirt out. Or rather, let us say it indicates either that the farmers' methods of caring for the milk are cleanly, or that the evidence of his uncleanliness has been strained out. It seems wise to advise dealers to prohibit the straining of milk at the farm. This may seem like radical doctrine, but a little thought discloses its wisdom. If the milk is unstrained, then the sediment test becomes of real value to the dealer in directing his attention to those dairymen who need attention from the inspector. Then, again, after the dirt once gains access to the milk, the straining simply removes that part of the dirt which is the least objectionable, namely, the insoluable part. If milk is poured over a strainer, the dirt remains upon the surface, and the milk which may be afterward poured through the strainer helps to carry through those soluble portions which may have at first escaped solution. This soluble portion, of course, contains most of the bacteria.

The removal of the dirt at the shipping station or the pasteurizing plant may be accomplished by straining as it is received, or since the straining of cold milk is difficult, it may be strained while hot from the heater and before it enters the holding apparatus. There is no objection to this, since whatever bacteria are carried through into the holder are destroyed during the holding process.

Clarification

The centrifugal clarifier is used to a large extent, and removes not only the dirt which is ordinarily seen in milk, but also the pus cells, blood cells, etc., which usually escape detection. The clarifier, as is well known, is made in a manner similar to a separator, the difference being that the cream is not separated from the milk, but the dirt, pus, etc., is thrown against the outer portions of the revolving shell, and is there collected. Of course, this material contains a great many bacteria, which are carried along with the heavier material. Tests, however, made by the plating method, do not show the great decrease in the number of bacteria which might be expected, and in many instances the apparent number is increased. This is probably due to the fact that the clumps of bacteria are broken up and so produce more colonies upon the plate than results when they remain bunched together in the original milk. One of the clarifiers is illustrated in Fig. 32.

One objection to the clarifier has been the

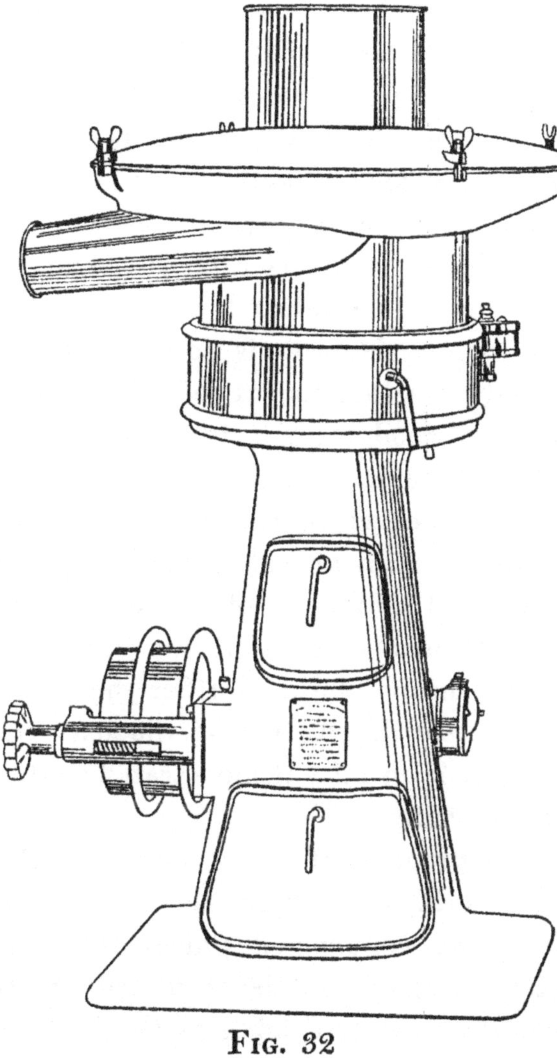

FIG. 32

tendency which there was to create foam, especially if the milk were cold. This foaming is due, of course, to the entrance of air into the milk during the process. Milk which has been previously pumped seems more likely to foam than when pumps have not been used. Machines are now made by which the air is excluded, and no foam, or almost none, is produced when they are operated, even when the milk is cold. It is generally held that the clarification of milk by a centrifugal machine is a good thing. It is claimed that the pus cells, leucocytes, etc., which are in the milk, tend to form a covering for the bacteria contained in the milk, to such an extent that they are somewhat protected and are not so readily destroyed by heat. It is claimed that the removal of this covering by means of the clarifier therefore assists in the complete destruction of the bacteria. This claim has not been fully demonstrated to the knowledge of the writer.

It is the practice of some dealers to strain the pasteuried milk as it enters the bottling machine. This is to be condemned unless the strainer is completely covered and is sterilized at the same time that the final sterilizing of the bottler occurs.

In fact, any straining after the milk leaves the holding apparatus is undesirable, and seems to be useless if the milk has been well cleaned before pasteurizing.

Coolers

The rapid cooling of heated milk has been held to be an important factor in the satisfactory pasteurizing of milk. It was claimed that the sudden chilling had a destructive effect upon the bacteria, which had been already weakened by the heating. Investigations made by Ayers and Johnson, however, seem to indicate that when the cooling is accomplished, during a period of two hours, there is no noticeable increase in bacterial growth.

There are three general types of milk coolers in use:

First: The open cooler or aerator. This consists of a series of tubes, made usually of copper and well tinned. These tubes are arranged one above another, and are so connected at the ends that cold water or brine is forced through the entire series. The water flow is into the lower tubes and upward through the series, so that the coldest water comes in contact with the coldest milk. This form

FIG. 33

of cooler is shown in Fig. 33. The cooler is usually arranged in sections. Through the upper section cold water, frequently well water, flows, while the lower section contains ice water or brine. This form of cooler is very effective in rapidly cooling the milk when it is made large enough to take care of the milk which is to be treated. They are made by various manufacturers in forms which vary somewhat in details. Some are so arranged that the regenerative principle is made use of. In most of these the hot milk from the holder is allowed to flow over pipes through which the incoming cold milk flows. This is a means of saving heat units, but such coolers are rather difficult to clean, and there is a danger that the pressure of the cold milk on the inside of the pipes will find openings through which the milk will be forced into the pasteurized milk on the outside. The contamination which results is not corrected by repasteurization. Some one may say that the same danger exists when the cooling medium on the inside of the pipes is water. This is true, but it must be remembered that the water pipes do not need to be taken apart for cleaning, while the milk pipes must of necessity be opened for cleaning daily, and the danger from loosened joints is therefore greater.

Another objection which has been urged to the use of the open cooler is that the milk is exposed to the air, and that there is great danger from air contamination. This is true when the air is full of dust or foul odors. In most milk plants, however, the air is so moist that little dust is flying about, and recent investigations by Ruehle and Kulp * of milk contamination by stable air indicate that even under unfavorable conditions the air contamination is not as serious a matter as we have supposed. Of course, milk coolers can be enclosed by tightly fitting metal covers, or they may be located in separate rooms or compartments, and so be protected from contact with air which is questionable. If a separate room is provided, suitable ventilation should be secured for carrying off steam, and the opening for the admission of air should be so protected that the incoming air will be filtered.

Since it is difficult to clean and sterilize metal covers for coolers, it seems more desirable that the milk be protected by enclosure in separate rooms.

The complete sterilization of this open type of cooler is a rather difficult matter. The reason for this is that the cooling surface cannot be subjected to the action of steam under pressure. The usual method employed is to direct live steam from a

*Bul. 409. N. Y. Agricul. Exp. Sta.

hose against the outer surface of the pipes, after the washing has been completed. It is well known that steam when relieved of its pressure rapidly cools, and a person can hold his hand a short distance from an open steam pipe with no inconvenience. It often happens that not all the parts of the cooler receive the benefit of even this reduced temperature. The tubes, of course, must be empty of the water usually contained in them before they are steamed.

It might be feasible to provide covers for these coolers which would be sufficiently tight to allow the pressure of the steam to be retained to some extent before it escaped into the outer air. Some covers are now made which are fairly tight.

The best way to sterilize this form of cooler is to have the tubes of sufficient strength so that steam may be admitted on the inside under pressure. In such case the sterilization is from the inside. If, while the pipes are hot from the enclosed steam, water is allowed to trickle over the outside, this is heated, and is itself converted into steam, which helps to cleanse the pipes.

Care should be taken to be assured that the steam which rises from the milk while being cooled does not become condensed upon any unclean surface, such as the ceiling or a metal covering over the cooler, and from thence fall back into the milk. If this occurs, very serious contamination may result.

Open pipe coolers should not be made very long in relation to their height. It can be seen that the water flowing through long pipes will become warmed on the inner surface of the tubes, while the central more rapidly flowing core may still be cool. More effective cooling is obtained where this milk flow is broken up by frequent turns as the water is returned from one pipe to another.

Second: The cooling may be accomplished in the same apparatus in which the milk is heated and held. This can be done when the tank system of pasteurization is employed, as illustrated in Figs. 12, 13, 14, 14-B, 15, 16. When cooling is so carried on, the cold water or brine takes the place of the hot water in the pipes, by means of which the heating was originally done. It is found in practice that when the same coil is used for the brine as well as for the hot water, there is danger that careless workmen may allow the brine and the water to become mixed. There is therefore a loss of brine. Some heating tanks are now made with double coils, one of which is used for heating and the other for cooling. If the cooling is done in tank pasteurizers there is the advantage that the milk comes in contact with no additional metal surface or other material after it is placed in the tank till it is drawn into the final containers.

The principal objection to the tank cooling is that usually a long time is required to reach the desired low temperature. As stated in the chapter on heaters, the time is frequently from an hour to an hour and a half. Large coils and a rapid circulation of brine will decrease this cooling time greatly.

In country plants at a distance from the point of milk consumption, the time of cooling is an important consideration from the dealers' point of view, since the time between the receipt of the milk from the farmer and the departure of the milk train is limited.

The change in temperature from the high heat necessary to pasteurize the milk to the cold required for rapid cooling produces a considerable strain upon the pipes of the coils, and the joints are apt to become weakened.

Third: The tubular cooler is being used to a considerable extent. In this form of cooler the pipes through which the hot milk flow are enclosed by other pipes, through which cold water is forced in a direction opposite to the flow of the milk. These coolers are exactly similar to the tubular heaters illustrated in Figs. 7, 10, and 11.

This form of cooler is effective and rapid; it has this advantage, that sterilization by steam under pressure can be made complete. There is the same difficulty in emptying the tubes which was noted in describing the tubular heater and holder, and whatever advantage there may be in exposing milk to the air in order to get rid of bad odors, etc., is absent from the tubular cooler.

The regenerative principle is made use of in the tubular heater and cooler, as already described, and there is little danger of the milk becoming contaminated by leakage of the water, since the joints of the water pipes do not come in contact with the joints of the milk pipes.

One of the most efficient forms of apparatus to use from the point of view of economy of heat units is probably that shown in Fig. 11. The great amount of piping, however, renders it difficult to keep clean.

Washing of Containers

It is of course important that containers in which pasteurized milk is placed shall be thoroughly cleaned and sterilized. There can be little advantage gained in the proper pasteurization of milk if it is to be immediately placed in containers which are not free from all dangerous bacteria. The relative importance which the cleaning of cans and bottles has in the bacterial content of milk has not been recognized to the extent which it deserves.

Cracks and open seams, especially in milk cans, may harbor particles of milk which are heavily infected with bacteria, which in many cases are of a type which are difficult to destroy by ordinary methods. Especially is this true if the cans or covers contain moisture, since this aids in the development of the bacterial life.

It is undoubtedly true that the methods employed in washing bottles are generally much more satisfactory than those in use in handling the cans and covers. Bottle-washing machines of various makes are on the market which perform excellent work. Some of these employ revolving brushes, which are automatically forced into the bottles, in which a strong washing solution has been placed. In others, the cleaning is accomplished by injecting under high pressure jets of washing solution into the inverted bottles. This type of machine is that which is in the most general use at the present time.

As usually constructed, this washing apparatus is a long machine, in which are placed several tanks for containing the washing and rinsing fluids. One tank contains a strong washing solution, another contains water for the first rinsing, and a third contains clean water for the final rinse. Over these tanks are placed plates having a surface area of about the same size as that of the top of a bottle case. These plates are perforated, and powerful pumps force the liquid in the tanks through these perforated plates in such a way that the liquid is forced upward into the bottles, which are inverted in cases over them. An automatic feed pushes the cases from plate to plate along a track made for the purpose. The bottles thus get first a cleaning with the fairly hot washing solution, then a rinsing with hot water, after which clean water is forced into them, which is near the boiling temperature, and sometimes a final jet of live steam is forced into the bottles. If the final rinse water is clean and hot, and the steam is applied long enough, the bottles leave the machine at a temperature of from 150° to 170°, and are reasonably free from bacteria.

Several points are to be borne in mind in using a machine of this kind. In the first place, no machine can be depended upon to fully clean an extremely dirty bottle, particularly those which are returned from the bottle exchanges and the dumps. Such bottles should first be washed by hand upon revolving brushes, where they can be clearly seen by the workman. They should then be washed again in the regular jet-washing machine. Another important point is that the perforations in the washing plates are liable to become clogged either with particles of dirt, or with rust, or it may be with deposits which are thrown down from water which is very hard. If these openings are stopped up it will mean that some bottles are not washed as it is intended they should be, and dealers may be unable to account for an unusually large number of dirty bottles which they find from time to time upon their routes.

Another important item is the temperature of the rinsing water. Unless this temperature is controlled by a thermostat, it may be easy to carelessly let the temperature drop below the sterilizing point. Workmen who have to take away the washed bottles from the machine and store them find that it is uncomfortable to handle bottles which are extremely hot, and these men may, therefore, allow the steam supply to the rinsing tank to be cut down to such an extent that the bottles do not get sterilized. A thermostat attached to this rinsing tank will automatically keep the temperature up to the desired point. If, in addition to this, a temperature recorder is also attached to this tank, a record will be kept which a dealer may keep on file as an office record. This will enable him to keep in closer touch with the work which is being done in his washing-room.

Another important matter is to make sure that the water which is used for the final rinse water is fresh, clean water, and that it is not used over and over in the final rinse tank. Most of the newer machines are so arranged that the water which is forced into the bottles for the final rinse, instead of falling back into the tank from which it is pumped, is carried back into the tank behind it. From here it can be properly used for the preliminary rinsing of the bottles. The use of fresh, clean water for final rinsing is somewhat expensive when a public water supply is used, and also a good deal of steam is required to heat it, but in order to obtain sterile bottles it is necessary. For heating this water use can be made of the exhaust steam from engines and pumps, and thus heating expenses may be cut down.

There is one form of automatic washing machine in which the bottles which are being treated are forced down into successive tanks of water, and the inverted bottles are pushed over fingers from which water is forced in a direction somewhat tangent to and also at right angles to the inserted finger. The water as it is forced into the bottle has therefore somewhat of a scouring action. This is of advantage in removing dirt which has adhered to the glass. In this machine, also, the water which is used for the final rinse is heated under pressure, and as a result, when it is forced into the bottle, its temperature is frequently above the boiling point. One feature of this machine is unique. The machine is so constructed that, when the final rinse water falls below a definite temperature, of say 180°, it refuses to operate, since the bottles will not be lifted from the rinse tank, but will remain submerged. The machine thus becomes its own thermostat. This effect is produced in the following manner: The platform upon which the cases are placed, and by means of which they are lowered into and lifted from the rinsing tank, is so counterbalanced that when the bottles are full of water the weight is too great for the case to be lifted. If the water which is forced into the bottles is hot enough so that as released from pressure it is converted into steam, then the water is forced out of the bottles, and they have sufficient buoyancy to enable the counterweight to lift the entire case out of the tank.

The older method used in washing bottles, even when large quantities were handled, was to provide means for soaking the dirty bottles, then to wash them by hand on revolving brushes, and finally to sterilize them in separate rooms, or in tanks into which live steam was discharged for a considerable time. This method is still in use to a considerable extent and is effective, but is more expensive than the machine method. This added expense is principally due to the amount of labor involved in the many handlings of the bottles which is necessary. Various soaking machines are in use. Some of them are so arranged that it is not necessary for the workmen to put their hands into the soaking solution, since an endless conveyor carries the bottles into and out of the soaking tank. The solution is made so strong that all foreign matter, even the bottle caps, is disintegrated. After being removed from the soaker, a rinsing by means of a machine which forces hot water into them renders them sterile.

Some method of thorough hand washing with subsequent careful sterilizing probably gives better results than does machine work, provided the men who do the washing are careful to see that no dirty bottles get by them. This is an expensive method of cleaning bottles.

Whatever method of washing is employed, careful attention should be given to proper inspection of the washed bottles. Probably no one thing can do more harm to a milkman's business than for his customers to frequently find bottles which have dirt adhering to the inside of them. Even though the dirt may have been sterilized by the steam treatment and be therefore harmless, the customer's sense of cleanliness is offended and the dealer may lose his trade.

A practice which has been adopted by some dealers commends itself to the careful milkman. This is the practice of restirilizing the bottles after they have been inspected and all dirty ones removed. Any contamination which may have resulted from the handling of the bottles by the workmen will thus be remedied.

After treatment, washed bottles should be stored either in an inverted position until they are used, or else they should be protected by suitable covers or by placing them in separate rooms where no infection can occur. If a cold-storage room is provided, for the washed bottles, there is a certain advantage. When it is remembered that the mass of glass in a milk bottle nearly equals the mass of the milk contained in it, it can be seen that if the bottle is warm, before it is filled, the temperature of the milk will be raised to a considerable extent. When empty bottles are stored at room temperature in the summer, they may have a temperature of 65° to 70° when the milk goes into them. If milk is cooled by the cooler no lower than 50°, it will, by coming in contact with the warm bottle, be warmed to a temperature of from 55° to 60°. Bacteria troubles may follow, especially if the cases of bottled milk are not immediately iced, or if they are placed in a cold room without ice being placed in contact with the bottles.

It is a known fact that milk either in cans or bottles will be very slow in cooling if the containers are placed in dry air, even though the temperature of the air is considerably below that of the milk. Milk containers, however, which have ice placed upon them, will become cooled in a short time.

Dealers and others have not realized to what an extent unclean cans and covers may, and in fact do, affect the bacterial content of the milk. Until comparatively recently, no effective can washing, sterilizing and drying machines have been placed on the market, and even now comparatively few milk dealers are using them to the extent to which their merit entitles them. It is usually found that in milk-handling plants, both in those where the milk is pasteurized and in those from which the milk is shipped in the raw state, the sterilizing is very inadequately done. Usually the cans are washed by hand with a brush, and are then placed over a single jet of rinsing water and then over a single jet of live steam. The time occupied in both rinsing and steaming is usually but a few seconds. It often happens that the can covers are not steamed at all, but are simply rinsed in water which is not over clean. When the cans which are to be returned to the patrons are washed, the steaming which the cans receive is so brief that it is almost negligible. This is not altogether the fault of the man who is washing the cans, for when milk is being delivered rapidly to the plant there is little time for long steaming. As a matter of fact, the cans require a more careful cleansing and sterilizing than do the bottles, since they are more likely to contain crevices and uneven surfaces than are the bottles. If for any reason it is not feasible for a dealer to install an automatic can washer, it is still possible to obtain satisfactory results. An apparatus which is comparatively inexpensive may be constructed by providing a long table or runway of iron plate on which there are installed two or three jets of water for rinsing cans, and also three or more jets of steam. Each jet should be provided with an automatic valve which will be opened by the weight or pressure of the can as it is placed over the jet. If flanges of iron or sections of gas pipe are so attached to the table that they will engage the flare of the can mouth as it is inverted over the table, it will be impossible to remove the can until it has travelled entire length of the table and has thus received the influx of water and steam from each of the jets. It is possible to so arrange the table that both cans and covers will be treated at the same time on the same table. This is accomplished by installing two sets of gas-pipe guides, one of them to hold the cans in place and the other of them to be just above the covers as they are pushed along the table, with the open end down. The cans referred to in this connection are the standard forty-quart cans used most largely in the East.

Various machines are on the market which are so arranged that the cans, after having been well steamed, are also dried by means of a blast of heated air. This drying is important for two reasons. First, because the moisture which may be left in the can renders the condition favorable for the growth of bacteria which have escaped destruction, and second, because cans which are not dry are much more likely to become rusty. Rusty cans are hard to clean, and frequently become a loss to the owner because they soon wear out and are discarded. One form of can rinser, steamer and dryer is rather ingeniously arranged. The exhaust steam, after having been used to operate a turbine-driven fan, is employed to sterilize the cans. The steam pipe which supplies the fan turbine passes in coils through an air chamber, and thus is used to heat the air which is forced by the fan into the cans after they have been sterilized. There is thus a very economical use of the steam, and the apparatus is also very compact.

Cans, after they have been treated, should be stored in an inverted position until they are used for containing milk. If, however, they are thoroughly dry, there seems to be no objection if the covers are tightly placed upon them and they are then stored without inverting.

In determining what test to apply in order to ascertain whether or not milk containers are sterile, or sufficiently so for practical purposes, it should be remembered that we are not dealing with laboratory conditions, and that therefore absolute sterility is not necessary, however desirable it may be from a theoretical standpoint. Our milk itself is not sterile, and if we can be assured that the containers are sufficiently clean so that they will not add any appreciable amount of bacteria to the milk, and that those which may be added are not of a pathogenic character, then health authorities should be satisfied. Remembering that a quart of milk contains about one liter of fluid, or 1,000 cubic centimeters, it seems reasonable to allow the washed bottles to contain 1,000 bacteria. This would add but one bacterium to each c.c. of milk contained in the filled bottle. In like manner, 40,000 bacteria in a washed can might add the same amount of bacterial contamination to the forty quarts of milk contained in the filled can. If none of these were of the coli group, it is probable that no danger would result. As a matter of fact, with methods of cleansing which may be easily secured, the bacterial content of the containers will be much less than the allowable amount above referred to.

Bottle-Filling

The filling of the bottles is more an economic problem than a sanitary one. Of course, the careful cleaning and sterilization of the apparatus is necessary, no matter what kind of filler is used. Any failure in this respect may cause the milk to gather up bacteria while it passes through the bottling machine. This may cause annoyance, at least, when tests are made which indicate a high bacterial content. If, however, the cleaning is carefully done, the items which most interest the milk men are the questions of expense and of efficiency in operation.

If a machine is operated by hand power, there is the question whether human power is cheaper than machine power. If automatically operated by outside power, either electric, steam or hydraulic, then the matter of bottle breakage, case damage, and loss of milk from broken bottles is an important consideration. A power-driven machine may or may not be slower in operation than hand power. The conditions surrounding each plant must be considered and decisions made in accordance with them.

The rotary filler and capper is coming into quite extensive use, with many features about it which appeal strongly to the milkman.

Bottle-Capping

It is important that milk, after having been pasteurized, shall not be recontaminated in any way. If containers are properly cleaned, the greatest danger of such reinfection lies in the practice of capping bottles by hand. When this method of capping is employed, the caps are frequently kept in open boxes beside the bottling machine, and they are sometimes kept in the pockets of the workmen who are capping the bottles. There is thus constant danger that the caps will become infected by dust in the air, by milk which is spattered upon them or by means of the hands of the persons who handle them. Men who handle bottle cases and at the same time work about machinery cannot easily keep the hands clean. The milk caps then suffer. If any germs of disease are about these employees then the danger becomes apparent.

Some form of capping machine is desirable. Several such are on the market, and many of them are doing satisfactory work. It is not necessary to describe the characteristics of each. Those are most desirable which fulfil the following conditions:

First: The caps used should be stored in sterile containers or tubes before they are inserted in the machine.

Second: A very small percentage of the bottles should be missed in capping.

Third: They should be so constructed that they can be readily taken apart for cleaning, and they should have no concealed pockets or spaces where milk is likely to collect and become sour or decomposed.

Fourth: They should be strongly constructed in order that they may stand without breakage or disarrangement, the rather rough usage to which they are subjected.

It is very desirable that milk bottles be sealed after they are filled. By sealing is meant that they be so protected that they cannot be opened without the fact being readily detected by the consumer. The desirability of this sealing process gets its force from two reasons:

First: To prevent the drivers or others from opening the bottles and adulterating the milk.

Instances have been known where the drivers of rival companies have opened the bottles belonging to their competitors, and have fouled the milk in order to cast discredit upon the milk delivered by the rival concern. Other drivers have taken cream from the tops of bottles, and have filled the bottles with water. The cream they have sold for their own profit. Bottles which were effectively sealed could not be treated in this way.

Second: To prevent possible contamination of the milk by means which are unintentional. The use of the usual paper plug cap leaves a space above the cap where dust and dirt and ice water from melting ice, etc., can collect. Unless this is carefully washed off before the bottle is opened, it is likely to fall into the milk. If a bottle of milk which when filled has a temperature of say 50° to 55° is then placed in a case and covered with ice, the milk is cooled to perhaps 35° to 40°. This causes the milk to contract. The air in the space above the milk also contracts, and a suction is thus created. The water which collects on the top of the cap is then drawn into the bottle around the edge of the cap, taking with it some of the dirt which may have collected there.

The drivers of retail delivery wagons who handle the reins, care for the horses and then in delivering milk grasp the bottles by the neck, are particularly likely to leave some dirt or contaminating material upon the lips of the bottles over which the milk will be poured. Serious infection may result. Caps which are real seals and which at the same time protect the neck and lips are on the market, but are rather expensive. They are used by dealers upon special grades of milk, such as Certified, Grade A, etc., but are as yet too costly to be used upon the lower or common grade of market milk. A great many attempts have been made to devise a safe seal which shall be protective and shall be inexpensive, but none have been very successful. A fortune awaits the man who can produce a satisfactory seal of this character.

Paper Bottles

For fifteen years or more various sorts of paper milk bottles have been upon the market. They have been made in various forms. Some were made of pulp which was formed into the shape of an ordinary bottle. Some have been crimped into a bottle form from a single sheet of heavy paper. Others have been made of heavy waxed paper, through which the cream was visible. Some have had straight parallel sides. They have been made round and square; have been filled from the top, and have been filled from the bottom, but the success which has attended them has not been very marked, especially in the East. There are some firms in the western part of the country where their use has been rather extensive and where they are said to be satisfactory.

The objections to their use has been—1st, the expense. They are more costly than are glass bottles, when it is considered that the glass bottle is used repeatedly, while the paper one is used but once.

2nd. Many of them are not strong enough to endure the heavy icing and rough handling to which they are subjected.

3rd. Customers have been accustomed to seeing the cream in the bottle, and object to bottles where the cream is out of sight.

4th. The constantly increasing cost of paper renders it questionable whether the equipment for using paper bottles may not prove to be available for a short time only, and its expense thus wasted.

5th. There is always uncertainty concerning the cleanliness of the bottles as delivered by the manufacturers for use, and also as to whether or not the material used in the paper may affect the milk.

There are undoubted advantages in the use of the paper container, among which are:

1st. That the expense of handling and transportation will be lessened, on account of the lesser weight of the paper.

2nd. The expense of washing returned bottles will be saved, since the paper container will be used but once.

3rd. The possible danger which exists through the return of bottles from the public dumps, and from houses in which there are communicable diseases will be lessened if the single service container is used.

4th. The danger from broken glass, which occasionally finds its way into the milk, will be obviated, and the wounds which are inflicted by broken bottles upon workmen who handle and wash them will be prevented.

Home Pasteurization

Many physicians who thoroughly believe that milk should be pasteurized for infant feeding, also recommend that this pasteurization be performed in the home, and that the pasteurized milk be fed as quickly as possible after the heating is completed. This method eliminates the possibility of recontamination which may occur when milk is handled under commercial conditions, and it also insures that those bacteria which are not destroyed during the pasteurization process will have little opportunity for growth and increase between the pasteurization of the milk and its consumption.

It is not the purpose of the writer to here describe the various methods employed for home pasteurization. Such information may be obtained from the pamphlets which are distributed by those responsible for maintaining the various infant-feeding stations located in our cities. Some of these stations are maintained by the municipalities and some by private individuals. Dr. Rosenau, in his book on "The Milk Question," discusses this phase of pasteurization quite fully.

Whatever method of home treatment is employed,

care should be taken that the heat is high enough and is maintained for a sufficient time to reach all parts of the milk, and so kill the pathogenic bacteria. The heating medium should not be hot enough to cook or change the chemical characteristics of the milk or to impart to it a disagreeable odor or taste.

CHAPTER VI

EFFICIENCY OF VARIOUS APPARATUS

During the writer's experience in the actual operation of pasteurizing plants, a great many samples of milk have been taken which have been examined for their bacterial content in order to determine the efficiency of the various pasteurizing plants in the destruction of bacteria. These results have been tabulated, showing the bacterial content before, during and after pasteurization. In these tables the kind of apparatus are indicated by figures only, no name of apparatus or of the manufacturers being given. The dealers at whose plants the tests were made are indicated as "A," "B," "C," etc. Thus at a plant operated by dealer "A," for instance, a heater illustrated by Fig. 00, a holder like Fig. 00, and a cooler like Fig. 00 may have been employed.

Lest the reader be misled as to the efficiency of certain definite forms of apparatus, and lest certain manufacturers be led to feel that results obtained have failed to give justice to the apparatus handled by them, the reader is warned that efficiency tests are sometimes misleading. It should be remembered that at least four factors enter into the causes which produce the effects indicated.

1st. The character of the apparatus itself.

2nd. The carefulness with which the plant is operated by those responsible for it.

3rd. Whether the apparatus is thoroughly cleaned and sterilized.

4th. The character of the milk treated.

It is, of course, understood by all those familiar with bacteriological work that these are factors of error in the results which can never be entirely overcome.

If apparatus is so constructed that the heating and holding and cooling are true to their indicated capabilities, then the results will be good—*if* the machine is properly operated and *if* it is clean. These factors being reliable, the results can be taken at their face value *if* the milk is of normal quality and does not contain, we will say, certain forms of bacterial life which are difficult to destroy.

When results are good, it undoubtedly means that both the apparatus and its operation are satisfactory. If the results are bad, it is not always easy to say whether the machine, its operation, or the milk is at fault.

In regard to the milk itself, there are many factors which must be taken into account when determining the efficiency of any apparatus. Estimates of machine efficiency based upon the percentage of bacteria destroyed are frequently misleading. It can be readily seen that if the bacterial content of the raw milk is high, any good pasteurizer will kill a large percentage of the germs in the milk. For instance, if the original milk contained 5,000,000 bacteria per c.c., and after being pasteurized contains 100,000 bacteria, there would have been a reduction of 98 per cent. of the bacteria in the original milk. Milk containing 100,000 bacteria per c.c. would not, however, be considered to be an entirely satisfactory product to place upon the market, especially if it contained organisms of the coli type.

Again, let us suppose that the original milk contained 10,000 per c.c., and that by means of the pasteurizing process this number was reduced to 1,000; such a milk would undoubtedly be a proper milk for consumption, although the percentage of reduction had been 90 per cent only. It is therefore plainly seen that the efficiency of pasteurizing machinery, when expressed in percentage reduction of bacteria only, must be taken with a great deal of reserve. It is necessary to know the character of the raw milk, and also to know if coli types are all destroyed. In speaking of the destruction of coli, it should be remembered that germs of this group are not necessarily all dangerous. In this connection they are used as indicators only. It is known that they are more difficult to destroy than are any of the pathogenic or disease-producing bacteria. If, therefore, the coli are all killed, it is safe to assume that no dangerous microbes remain alive. While, if the coli are present in the finished product, there is a possibility that disease germs are also present. It is known pretty definitely that if milk is heated to 142-145° for thirty minutes, all coli will be destroyed.

The following tables are compiled from very many tests made. When these tests were made there were three or four samples of milk taken at each point of examination; for instance, four samples of raw milk were taken at the same time. When these were examined, the average of all the counts was determined. In the same way there were four samples taken at the outlet of the heater, the holder, and so on. In the tables herewith compiled, only the average counts are indicated, in order to avoid any bewildering mass of figures. It should be therefore borne in mind that bacteria counts here indicated are averages.

In figuring out the percentage of reduction, the relation between the bacterial content of the raw milk and the milk as it leaves the holder is considered. It will frequently be found that the number of bacteria in the milk leaving the cooler is higher than when it leaves the holding apparatus, and that this number is again increased when the milk enters the bottles and cans. Usually unclean coolers, bottling machines and bottles are responsible for this increase.

Not all of the types of apparatus which are illustrated and described in this book have been tested by the writer as to their efficiency in the destruction of bacteria, and any omissions which may be observed in the tables are not due to any intention on the part of the writer to slight any of them, or to any desire to call attention more forcibly to one kind of machine than to another. It simply means that the results of tests are not available.

PASTEURIZING PLANT OF DEALER "A"

Heater..................................Fig. 4 (modified)
Holder..................................Fig. 20
Cooler..................................Fig. 11

1st Series	Raw Milk	From Heater	From Holder	From Cooler	In Bottles	Percentage of Reduction
Bact. per c.c......	8,500	800	634	.....	350	92.55%
Temp..........	44°	138°	139°	.....	47°	
2nd Series						
Bact. per c.c......	213,334	12,734	1,150	2,800	5,700	94.57%
Temp..........	44°	137°	139°	.....	.....	
3rd Series						
Bact. per c.c......	1,650,000	51,250	1,250	2,550	9,375	99.93%
Temp..........	50°	142°	142°	48°	49°	

PASTEURIZING PLANT OF DEALER "B"

Heater..Fig. 3
Holder..Fig. 20
Cooler..Fig. 33

1st Series	Raw Milk	From Heater	From Holder	From Cooler	In Bottles	Percentage of Reduction
Bact. per c.c.	126,500	5,350	200	11,600	49,210	99.42%
Temp.	40°	145°	139°	38°	40°	
2nd Series						
Bact. per c.c.	97,666	7,400	1,266	800	3,466	98.71%
Temp.	40°	147°	144°	40°	42°	
3rd Series						
Bact. per c.c.	15,000	. .	2,850	150	5,300	81%
Temp.	48°	. .	140°	46°	47°	

PASTEURIZING PLANT OF DEALER "C"

Heater.......................Fig. 4
Holder.......................Fig. 20
Cooler.......................Fig. 11

1st Series	Raw Milk	From Heater	From Holder	From Cooler	In Bottles	Percentage of Reduction
Bact. per c.c......	102,500	6,300	600	150	500	99.12%
Temp............	46°	147°	140°	40°	42°	
2nd Series						
Bact. per c.c......	247,500	20,600	2,300	460	4,350	99.07%
Temp............	46°	145°	143°	40°	44°	

PASTEURIZING PLANT OF DEALER "D"

Heater........................ ....Fig. 3
Holder........................ ....Fig. 24 (1 tank)
Cooler........................ ....Fig. 33

1st Series	Raw Milk	From Heater	From Holder	From Cooler	In Bottles	Percentage of Reduction
Bact. per c.c.... ..	4,136,000	587,600	132,300	6,400	71,600	96.56%
Temp.......... ..	46°	140°	128°	40°	45°	
2nd Series						
Bact. per c.c.... ..	9,266,000	110,800	2,300	5,200	79,800	99.98%
Temp.......... ..	46°	143°	143°	42°	45°	

PASTEURIZING PLANT OF DEALER "E"

Heater.................................Fig. 3
Holder.................................Fig. 24
Cooler.................................Fig. 33

Holding time, 30 minutes

1st Series	Raw Milk	From Heater	From Holder	From Cooler	In Bottles	Percentage of Reduction
Bact. per c.c......	5,325,000	126,200	10,000	7,000	16,250	99.82%
Temp............	48°	150°	148°	41°	45°	
2nd Series						
Bact. per c.c......	7,400,000	16,030	7,000	5,800	10,300	99.97%
Temp............	46°	150°	147°	42°	45°	
3rd Series						
Bact. per c.c....	174,000	12,800	800	3,600	5,400	99.54%
Temp..........	47°	151°	148°	43°	45°	
4th Series					Cans	
Bact. per c.c...	365,000	1,000	100	100	3,450	99.98%
Temp.........	40°	162°	160°	50°	50°	

PASTEURIZING PLANT OF DEALER "F"

Heater .... Fig. 3
Holder .... Fig. 24
Cooler .... Fig. 7

Holding time, 35 minutes

1st Series	Raw Milk	From Heater	From Holder	From Cooler	In Bottles	Percentage of Reduction
Bact. per c.c.	603,000	.....	3,400	.....	2,160	94.27%
Temp.	40°	.....	140°	.....	45°	
2nd Series						
Bact. per c.c.	5,238,000	.....	4,400	.....	34,000	99.98%
Temp.	41°	.....	141°	.....	46°	
3rd Series						
Bact. per c.c.	168,000	.....	375	2,500	2,050	99.77%
Temp.	40°	.....	142°	40°	42°	
4th Series				Cans		
Bact. per c.c.	259,000	1,950	100	38,000	18,700	99.96%
Temp.	72°	174°	163°	43°	.....	
5th Series						
Bact. per c.c.	135,000	1,250	−100	8,150	126,000	99.97%
Temp.	68°	148°	144°	42°	44°	

PASTEURIZING PLANT OF DEALER "G"

Heater.............................Fig. 11
Holder.............................Fig. 21
dGr.............................Fig. 11

1st Series	Raw Milk	From Heater	From Holder	From Cooler	In Bottles	Percentage of Reduction
Bact. per c.c......	3,316,000	12,233	700	2,633	4,000	99.99+%
Temp............	45°	143°	150°	46°	52°	
2nd Series						
Bact. per c.c......	5,133,300	128,250	1,167	400	9,366	99.99+%
Temp............	41°	143°	147°	47°	52°	

PASTEURIZING PLANT OF DEALER "H"

Heater . . . . Fig. 7
Holder . . . . Fig. 27
Cooler . . . . Fig. 7

First milk through apparatus for day

1st Series	Raw Milk	From Heater	From Holder	From Cooler	In Bottles	Percentage of Reduction
Bact. per c.c.	120,000	100,000	8,200	20,000	84,000	93.18%
Temp.	42°	140°	136°	44°	44°	
2nd Series						
Bact. per c.c.	490,000	170,000	450	500	10,000	99.91%
Temp.	58°	142°	144°	50°	56°	
3rd Series						
Bact. per c.c.	7,153,000	. . . .	1,113	. . . . .	5,700	99.99%
Temp.	42°	. . . .	144°	48°	48°	
4th Series						
Bact. per c.c. . . .	58,000	. . . .	400	. . . . .	7,825	99.31%
Temp.	42°	. . . .	144°	48°	48°	

PASTEURIZING PLANT OF DEALER "I"

Heater..............................Fig. 7
Holder..............................Fig. 24
Cooler..............................Fig. 7

Holding time, 20 minutes

	Raw Milk	From Heater	From Holder	From Cooler	In Bottles	Percentage of Reduction
1st Series						
Bact. per c.c.	253,000	27,000	17,000	5,300	351,000	93.29%
Temp.	45°	140°	139°	46°	48°	
2nd Series						
Bact. per c.c.	2,740,000	49,300	3,200	1,400	51,100	99.98%
Temp.	46°	140°	39°	47°	49°	

First milk passing through tank for day
Holding time, 20 minutes

	Raw Milk	From Heater	From Holder	From Cooler	In Bottles	Percentage of Reduction
3rd Series						
Bact. per c.c.	232,500	2,900	37,500	1,750	400	83.45%
Temp.	38°	145°	140°	38°	40°	

Milk passing through tank later during operation

	Raw Milk	From Heater	From Holder	From Cooler	In Bottles	Percentage of Reduction
4th Series						
Bact. per c.c.	295,000	57,950	4,400	14,500	9,550	98.51%
Temp.	38°	140°	138°	40°	44°	

PASTEURIZING PLANT OF DEALER "J"

Heater.................................Fig. 7
Holder.................................Fig. 24
Cooler.................................Fig. 7

1st Series	Raw Milk	From Heater	From Holder	From Cooler	In Bottles	Percentage of Reduction
Bact. per c.c....	120,000	16,800	5,300	15,475	2,750	95.70%
Temp.........	40°	150°	120°	45°	46°	
2nd Series						
Bact. per c.c....	33,250	12,700	567	100	550	98.20%
Temp.........	45°	144°	143°	44°	52°	

PASTEURIZING PLANT OF DEALER "K"

Heater..............................Fig. 1
Holder..............................Fig. 26
Cooler..............................Fig. 33 (regenerative)

Holding time, 20 minutes

1st Series	Raw Milk	From Heater	From Holder	From Cooler	In Bottles	Percentage of Reduction
Bact. per c.c.	48,000	3,580	160	240	283	96.67%
Temp.	45°	148°	143°	41°	43°	
2nd Series						
Bact. per c.c.	7,583,000	1,313,300	1,400	2,660	6,766	99.98%
Temp.	48°	147°	142°	43°	46°	
Holding time, 10 minutes						
3rd Series						
Bact. per c.c.	7,533,000	920,000	1,367	934	2,067	99.98%
Temp.	47°	152°	149°	45°	47°	
Holding time, 20 minutes						
4th Series						
Bact. per c.c.	823,000	17,500	500	−100	−200	99.94%
Temp.	46°	148°	146°	33°	40°	

PASTEURIZING PLANT OF DEALER "L"

Heater..............................Fig. 1
Holder..............................Fig. 26
Cooler..............................Fig. 33 (regenerative)

Holding time, 20 minutes

1st Series	Raw Milk	From Heater	From Holder	From Cooler	In Bottles	Percentage of Reduction
Bact. per c.c.	3,540,000	22,400	2,200	900	16,000	99.94%
Temp.	108°	145°	140°	54°	54°	
2nd Series						
Bact. per c.c.	5,800,000	67,600	23,000	20,000	34,600	99.61%
Temp.	100°	146°	142°	53°	54°	
3rd Series						
Bact. per c.c.	11,062,500	258,200	95,000	90,800	142,000	99.13%
Temp.	103°	145°	140°	53°	54°	
4th Series						
Bact. per c.c.	610,000	3,400	4,900	5,100	966	92.97%
Temp.	104°	147°	146°	44°	44°	

PASTEURIZING PLANT OF DEALER "M"

Heater . . . . Fig. 8
Holder . . . . Fig. 24 (1 tank)
Cooler . . . . Fig. 33

Holding time, 15 minutes

1st Series	Raw Milk	From Heater	From Holder	From Cooler	In Bottles	Percentage of Reduction
Bact. per c.c.	49,200	5,500	6,300	1,375	3,900	87.20%
Temp.	56°	140°	137°	49°	50°	
2nd Series						
Bact. per c.c.	507,000	14,600	103,700	633	1,700	77.55%
Temp.	48°	139°	136°	50°	50°	

PASTEURIZING PLANT OF DEALER "N"

Heater........................ ....Fig. 5
Holder........................ ....Fig. 22
Cooler........................ ....Fig. 5

1st Series	Raw Milk	From Heater	From Holder	From Cooler	In Bottles	Percentage of Reduction
Bact. per c.c...	546,000	69,000	90,000	33,000	98,000	98.36%
Temp.........	49°	150°	132°	34°	48°	
2nd Series						
Bact. per c.c......	1,716,000	113,500	116,000	84,000	64,000	93.24%
Temp............	50°	145°	141°	40°	50°	
3rd Series						
Bact. per c.c....	333,000	14,000	3,000	4,000	66,000	99.10%
Temp..........	38°	147°	142°	36°	40°	
4th Series						
Bact. per c.c....	783,000	51,000	4,000	1,600	3,600	99.49%
Temp..........	38°	146°	142°	36°	40°	

PASTEURIZING PLANT OF DEALER "O"

Heater..............................Fig. 5
Holder..............................Fig. 24 (1 tank)
Cooler..............................Fig. 5

1st Series	Raw Milk	From Heater	From Holder	From Cooler	In Bottles	Percentage of Reduction
Bact. per c.c....	270,000	.....	1,000	.....	.....	99.63%
Temp..........	45°	150°	148°	.....	.....	
2nd Series						
Bact. per c.c....	201,000	.....	1,500	....	.....	99.3%
Temp.........	43°	145°	143°	....	.....	
3rd Series						
Bact. per c.c......	78,000	8,600	556	100	433	99.64%
Temp............	40°	145°	137°	42°	43°	
4th Series						
Bact. per c.c......	246,000	45,000	100	160	666	99.96+%
Temp............	37°	145°	138°	45°	46°	

PASTEURIZING PLANT OF DEALER "P"

Heater . . . . Fig. 13 (3 tank)
Holder . . . . Fig. 13
Cooler . . . . Fig. 00 (disk tank not shown)

1st Series	Raw Milk	From Heater	From Holder	From Cooler	In Bottles	Percentage of Reduction
Bact. per c.c. . . .	993,000	202,000	85,000	170,000	256,000	82.45%
Temp.	60°	142°	142°	37°	47°	
2nd Series						
Bact. per c.c. . . .	292,000	17,000	2,500	4,000	1,000	99.14%
Temp.	61°	142°	142°	38°	46°	
3rd Series						
Bact. per c.c.	2,350,000	123,300	1,000	8,000	10,600	99.96%
Temp.	62°	143°	37°	47°	. . . . .	
4th Series						
Bact. per c.c.	1,093,000	106,000	12,000	22,400	19,000	98.81%
Temp.	72°	143°	146°	40°	50°	

PASTEURIZING PLANT OF DEALER "R"

Heater..............................Fig. 13 (3 tank)
Holder..............................Fig. 13 (3 tank)
Cooler..............................Fig. (Disk cooler not shown)

Holding time, 20 minutes.

1st Series	Raw Milk	From Heater	From Holder	From Cooler	In Bottles	Percentage of Reduction
Bact. per c.c.	33,000	4,300	260	160	450	99.22%
Temp.	40°	142°	130°	33°	41°	
2nd Series						
Bact. per c.c.	125,000	80,000	300	434	334	99.77%
Temp.	38°	140°	136°	34°	40°	

PASTEURIZING PLANT OF DEALER "S"

Heater.................................Fig. 10
Holder.................................Fig. 23
Cooler.................................Fig. 10

Holding time, 30 minutes

1st Series	Raw Milk	From Heater	From Holder	From Cooler	In Bottles	Percentage of Reduction
Bact. per c.c....	484,000	23,300	300	15,400	7,100	99.94%
Temp..........	38°	146°	146°	35°	35°	
2nd Series						
Bact. per c.c....	658,000	4,000	233	666	800	99.98%
Temp..........	38°	146°	146°	35°	35°	

CHAPTER VII

CHANGES IN THE CREAM LINE DUE TO THE PASTEURIZATION OF MILK

Dealers who have had experience in the pasteurization of milk know that from time to time they have trouble with what is known in the trade as the "cream line" upon bottled milk. This may mean that the line of demarcation between the cream in the bottle and the skim milk just below it is indistinct and faint, or it may mean that no cream at all is visible and the contents of the bottle has the same color all the way through. Again, it may mean that the apparent amount of cream upon the milk is lessened. Either one of these conditions is a reason for complaint on the part of the customers. Dealers are therefore anxious to obtain as great a volume of cream as possible upon the milk sold by them, and are also anxious to get the line of division as distinct as possible. Dealers are sometimes at a loss to know why milk will at one time show a good cream line, and at another time, when the conditions are apparently the same, the cream will be disturbed.

In 1914, when the writer was connected with the New York City Department of Health, the Board of Health adopted certain modifications in the regulations which had been in force concerning the pasteurization of milk. Among these changes were different rules concerning the temperature to which milk must be heated and the length of time for which it must be held at this temperature if it was to be officially recognized as pasteurized milk.

Prior to that time, the rules allowed a somewhat wide variation in the heating temperatures, and a corresponding variation in the length of time for which the milk should be held. Thus milk which was heated to 140° must be held at least twenty minutes, while milk heated to 158° need be held but three minutes. Between these extremes other temperatures and holding times were allowed. This sliding scale, as it might be called, allowed such a latitude that milkmen who sold milk in cans only, and were thus not much concerned with the cream line, could heat the milk to a high temperature and hold it for a short time. On the other hand, those dealers who sold bottled milk, and who were desirous to obtain as great a volume of cream as possible, could and usually did heat the milk to a low temperature and hold it for a comparatively long time.

A desire for uniformity of methods in the treatment of milk led the authorities to adopt regulations, in accordance with which one standard for pasteurization only was applied, in all cases when milk was heated for market use. The rules thus adopted required all pasteurized milk to be heated to 145° for thirty minutes.

Dealers who endeavored to comply with this rule complained that they found the cream line was affected, and that they could not comply with the public requirements without injury to their business.

In an effort to ascertain to what extent the strict enforcement of the regulations would affect the commercial quality of the milk, the writer made an investigation.

This investigation was not made with the intention to determine what form of apparatus would do the best work, nor to make any extended or exhaustive tests of any one form of apparatus under varying surroundings and conditions. The object was to see if it was possible and commercially practicable to heat milk to 145° and hold it for thirty minutes with the machines in use in the city without seriously affecting the cream line. Any results and data secured do not warrant the conclusions that the experiments indicate the essential superiority of any of the apparatus described.

A little thought on the part of the reader will make it clear that so many factors enter into the problem that such conclusions would be unwarranted.

In the same way that percentages of bacteria reduction mean little unless we possess all the data connected with the handling of the milk, so likewise percentages of cream upon pasteurized milk are not alone of ultimate value as showing that one apparatus is more satisfactory than another.

A series of tests was made at several of the pasteurizing plants in New York City at which different forms of apparatus were being used, the milk in these various plants being subjected to different degrees of temperature for varying lengths of time. Samples of milk were taken to determine the amount of cream which appeared upon the bottled product. At the same time, the conditions surrounding the handling of the milk were noted in order to determine if other factors aside from the temperature and holding time had any effect upon the cream line.

In making the tests, the samples of milk taken were set in standard cream gauge glasses and the actual percentage of cream showing upon the milk as indicated thereon was carefully noted. These tests were made during the early part of December, 1914.

First Test.—The first test was made at a plant in which the pasteurizing apparatus consisted of three tanks of the type shown in Fig. 13. Each tank had a capacity of about forty cans of milk, and after being filled the milk was heated by means of a revolving coil of pipe extending through the tank, through which coil hot water and steam were forced. After the milk reached the proper temperature, the steam was shut off in this coil and the milk was held at the same temperature for the desired length of time. It was then discharged into a cooler consisting of a tank in which were revolving disks, which is not here illustrated, and immediately cooled. Since it required considerable time for the tank to be fully emptied after the discharge was commenced, it is evident that a part of the milk was held in the tank for a considerably longer time than was that which was first discharged from the tank. In making the test, one tankful of milk was heated to a temperature of 142° and was held thereat for about thirty minutes. The exact time of holding was not ascertained on account of the fact that the milk was being discharged from the tank at the time the inspection began. A sample was taken from this milk and placed in a cream gauge, which was then packed in ice and allowed to stand for four hours, at the end of which time 15 per cent. of cream appeared upon the cream gauge.

A second sample was taken from one of the tanks in which the milk had been heated to a temperature of 146°, and held thereat for thirty-four and a half minutes. The sample here taken was allowed to stand packed in ice for four hours, at the end of which time 13 per cent. of cream appeared upon the gauge. A third sample was taken from the same tank, but it was taken from that milk which was last discharged from the tank. This milk, therefore, had been held in the tank for one hour and twenty-three minutes. This sample was also set in ice for four hours and showed 8 per cent. of cream upon the gauge.

A fourth sample was taken from a second tank in which the milk was heated to a temperature of 146° and held thereat for exactly thirty minutes. This milk showed 15 per cent. of cream upon the gauge.

The method of handling the milk at this plant should be noted. Milk was poured from the cans into a dumping tank on the ground floor. It was then pumped to a small tank upon the third floor of the building, thence flowed through a centrifugal clarifier, such as shown in Fig. 32, and then flowed directly into the pasteurizing tanks.

Second Test.—A test was also made in another milk plant in which the same form of pasteurizing apparatus was installed. In this plant, the milk was pumped from the receiving vat on the gound floor into a tank upon the second floor, from which the milk flowed through a clarifier and was then discharged into a mixing vat, from which it flowed into the various pasteurizing tanks. The operation of this plant was similar to that in the plant previously described, except that, in the first instance, the water which was forced through the heating coil of the tank was heated by means of a steam jet, injecting steam directly into the water pipe, whereas, in the second instance, water was heated in a tank outside of the pasteurizing tank and was pumped, after the heating, directly into the heating coil, it being the belief that if the water was heated in this manner there was less danger of the milk becoming scorched on account of the water coil having been heated to a temperature which was excessive

At this second plant the first sample was taken from the raw milk before it had passed through the clarifier. This sample was allowed to stand in ice for eleven hours, at the end of which time 19½ per cent. of cream appeared upon the gauge. A Babcock test was made of this milk and it was found to contain 3.8 per cent. of butter fat. A second sample was taken, which was as nearly as possible from the same milk, after it had passed through the clarifier. This sample, after standing for eleven hours, showed 17 per cent. of cream on the gauge. This milk also contained 3.8 per cent. of butter fat. A third sample was taken from one of the pasteurizing tanks in which the milk was heated to 145° and was held in the tank for thirty-three minutes, at which time the temperature was found to be 142½° F. This sample, upon being set, showed 12 per cent. of cream in the cream gauge and contained 3.8 per cent. of butter fat. The fourth sample was taken from the same tank of milk, but when the tank was nearly empty. The milk had been standing in this tank for one hour and twenty minutes. It had been, however, cooled to a temperature of 60° F. by means of cold water pumped through the revolving coil. A sample of this milk, upon setting, showed 15 per cent. of cream in the cream gauge and contained 3.8 per cent. of butter fat.

It was thought possible that the speed of cooling of milk might have an effect upon the cream line. Therefore two samples of milk were taken, one of which had a temperature of 135° F. This sample was allowed to stand un-iced until cooled by air contact to a temperature of about 40° F. The second sample of milk was taken from the same batch of heated milk, but this sample was taken after the milk had been quickly cooled in the cooling apparatus to a temperature of 32° F. After setting, these two samples showed no appreciable difference in the cream content. This indicated that the speed of cooling had little or no effect upon the rising of the cream in the milk.

The tests made at these two plants indicated that it was entirely possible to heat milk to 145° F. and hold it for thirty minutes without materially affecting the volume of the cream contained in the bottled milk, and that milk heated to 145° F. showed the same amount of cream as that which is heated to 142° F.

THIRD TEST.—A test was made at another plant where the same form of apparatus for pasteurizing was in use. During this test a sample was taken from a tank in which the milk was heated to a temperature of 141° and held there for sixty minutes. This sample, after standing a sufficient length of time, showed a cream content of 15 per cent. A second sample was taken of raw milk which had already been passed through the clarifier, and this milk was found to contain 15 per cent. of cream. A third sample was taken from a tank in which the milk had been heated to 145° and held thereat for sixty minutes. This milk, after standing, showed no cream at all upon the cream gauge.

The results here obtained did not correspond with those obtained at the two other plants where the same apparatus was in use. No apparent reason could be seen for the variation. Milk at this plant, as received from the patrons, was allowed to flow from the weigh vat into the mixing vat, from which it was pumped to a vat upon the second floor, thence it flowed into a clarifier, from which it was discharged into a large mixing vat on the same floor. It then flowed by gravity to the pasteurizing vats on the first floor.

FOURTH TEST.—A further test was made at another pasteurizing plant, in which the milk was pasteurized by means of a heater and holder, shown in Figs. 5 and 22. In this heater the milk was forced between two concentric cylinders, in each of which hot water was contained, having a temperature of 159°. The milk while being heated was kept in agitation by the rapid revolution of one of the cylinders above referred to. After heating, the milk flowed by gravity into a series of insulated tanks, in each of which it was held for twenty-four minutes. On account of the fact, however, that milk as discharged from the heater flowed first into a tank from which the various holding tanks were filled, and on account of the added fact that when these tanks were discharging the milk, it flowed into a second tank before being conveyed to the cooling apparatus, it is probable that the milk was actually held at the highest temperature for a period of at least twenty-seven minutes.

A sample was taken from the raw milk before being clarified. This milk, upon setting, was found to contain 15 per cent. of cream and to have a fat content of 3.9 per cent. A second sample was taken from milk which had been heated to 145° F., and, after holding, was set and found to contain 10½ per cent. cream and 3.9 per cent. butter fat. A third sample was taken from milk which was heated to 144°, and was found to contain 8 per cent. of cream and 3.9 per cent. butter fat. A fourth sample was taken from milk which was heated to 142°, and was found to have a cream content of 8 per cent. and 3.9 per cent. butter fat.

These results indicated that the milk which was heated to 145° contained a greater percentage of cream, by volume, than that which was heated to 142°. The reason for this was not apparent. The milk pasteurized at this plant was received from the railroad in cans which had been shipped from the country. The milk was dumped from these cans into a large dump vat, from which it was pumped to a receiving vat upon the second floor and was thence allowed to flow through a clarifier, from which it was discharged into a large mixing vat and thence flowed directly to the milk heater.

Fifth Test.—This test was made at a plant at which the milk was pasteurized by means of a heater somewhat similar to that shown in Fig. 3. The holder was like Fig. 21. In this heater the milk is forced between two water-heated cylinders, in one of which the water was at a temperature of 165°, and in the other of which the temperature of the water was nearly at the boiling point. The milk, while being heated, was kept in agitation by means of a revolving arm, which was so arranged that it moved between the two heating surfaces. The milk, after heating, was held in a series of eight water-jacketed tanks, the water surrounding the tanks being kept at a temperature of from 145° to 150° F. The milk, while in this holding tank, was kept in agitation by means of revolving metal stirrers.

A sample was first taken from one of the holding tanks in which the milk was heated to a temperature of 143° F. and held there for thirty minutes. This sample, when set, was found to have 11 per cent. of cream in volume, and the butter fat content was 3.7 per cent. A second sample was taken from a holding tank in which milk had been heated to a temperature of 147° and held for thirty minutes, the temperature of the milk at its discharge being 146°. This milk showed 4 per cent. of cream and a fat content of 3.8 per cent. A third sample was taken from a tank in which milk was heated to 145° and held for thirty minutes. This milk, when set, showed 7 per cent. of cream and 3.7 per cent. butter fat. A fourth sample was taken from milk which had been heated to 146° and held for thirty minutes. This milk showed 3 per cent. of cream and 3.9 per cent. butter fat.

These results indicate that with this form of apparatus the cream line was seriously interfered with, even at a temperature as low as 143°. The milk, as received in this plant, was poured from cans into a receiving tank upon the upper floor of the building from which it flowed to an apparatus on the floor below in which it was preheated to 60° F. It then passed through centrifugal clarifiers from which it was discharged into a large tank upon the receiving floor. From this tank it flowed by gravity to the pasteurizers on the floor below.

SIXTH TEST.—This test was made at a plant in which the milk was heated by means of an apparatus shown in Fig. 11, where the milk is heated by means of what is known as a Multitube Heater. In this form of heater a series of large tubes are so arranged that several smaller tubes extend through them. The milk is forced through the inner tubes, while hot water is forced through the large surrounding tubes. After heating, the milk is discharged into a series of eight holding tanks similar to those described in test No. 5. The water which is used to heat the milk in this apparatus is itself heated in a tank outside of the milk heater and is kept at a temperature varying between 144° and 163°, the temperature rarely exceeding 154°. From this tank the water is pumped to the milk-heating apparatus.

The conditions at this plant were apparently favorable for securing good results. The fact that the temperature of the heating water was carefully controlled, made it impossible to superheat or scorch any of the milk. The first sample of milk was taken from raw milk before it was clarified. This sample, when set, showed 13½ per cent. of cream and 3.4 per cent. butter fat. The second sample was taken from one of the holding tanks in which the milk was heated to a temperature of 146° and held thereafter for thirty minutes. This sample, when set, showed no cream whatever on the cream gauge, the fat being evidently evenly distributed throughout the entire body of the milk. This milk contained 3.5 per cent. of butter fat. A third sample was taken from one of the holding tanks in which the milk was heated to 143° F. and held for thirty minutes. This milk showed, upon standing, 10 per cent. of cream and 3.7 per cent. butter fat. Sample No. 4 was taken from a tank in which the milk was heated to 146° F. and held for thirty minutes. This sample showed 5 per cent. of cream and 3.35 per cent. butter fat.

The results here obtained indicate that with this apparatus if the milk was heated above 143° the cream would be seriously affected.

Seventh Test.—A test was made at still another plant in which the milk was heated by means of the pasteurizer shown in Fig. 9. In this apparatus the milk is forced through a series of pipes which are enclosed in a large chamber which is filled with hot water. The apparatus somewhat resembles the tubular boiler. After heating, the milk is held in three upright cylindrical Park holding tanks, the milk passing through the entire series. In this holding apparatus tests previously made indicated that milk is held for from thirty-five to forty minutes.

In making this test, the first sample was taken from the raw milk. This sample, when set, showed 14 per cent. of cream and 3.8 per cent. butter fat. A second sample was taken from milk which was heated to a temperature of 143° and held for from thirty-five to forty minutes. This sample, when set, showed 14 per cent. of cream and 3.8 per cent. butter fat. In order to vary the experiment the temperature of the milk entering the holder was then raised to 145°, at which temperature the milk was maintained until the process of pasteurizing was completed for the day. The milk in the filled holding tanks was allowed to stand for thirty minutes and a sample was taken from that which was being discharged from the holding tank, which showed, on standing, 14 per cent. of cream and 3.85 per cent. butter fat. A fourth sample was taken from the milk which was last discharged from the holder. This milk had been in the holder for exactly sixty minutes. Upon standing, this milk showed 9 per cent. of cream and 3.7 per cent. butter fat.

These results indicated that milk could be heated to 145° and held for thirty minutes without any injurious effect upon the cream line being produced, but that if the milk was held for an hour, the cream line was considerably affected.

Eighth Test.—A further test was made at a pasteurizing plant in which the milk was heated by means of the heater shown in Fig. 7, and held in the holder illustrated in Fig. 27. In this apparatus the milk is forced through a series of pipes which are themselves enclosed in a larger pipe through which hot water is forced. The holding apparatus consists of a series of large tubes through which the milk is allowed to flow. It requires about thirty-five minutes for the milk to be discharged through the holding tubes.

A sample was taken from the raw milk, before clarifying. This was found to contain 15½ per cent. of cream and 3.5 per cent. butter fat. A second sample was taken from milk which had been clarified, and this was found to contain 12 per cent. cream and 3.6 per cent. butter fat, there being an apparent reduction of 3½ per cent. of cream volume due to the clarifying process. A third sample was taken from milk leaving the holder at a temperature of 142°. This was found to contain 8 per cent. cream and 3.7 per cent. butter fat. A fourth sample was taken from milk leaving the holder at a temperature of 145°. This showed 7 per cent. of cream and 3.6 per cent. butter fat.

The results here obtained indicated that milk which is heated to 145° shows about the same amount of cream, by volume, as that which is heated to 142°, but the milk heated to either temperature showed but little more than half the amount of cream which was found in the raw milk.

The milk at this plant was received in the dumping tank on the first floor, from which it was pumped to a tank upon the pasteurizing floor, and thence flowed through a centrifugal clarifier into the mixing tank. From this tank it was pumped through the heater and holder, and after being discharged from the holder was again pumped through the cooler.

Summary of Results

All the tests made indicate that when milk is cleaned by a centrifugal clarifier, the volume of cream in the milk suffers a reduction of 2 per cent. to 3 per cent.

The number of tests made was probably not sufficient to warrant the drawing of absolutely definite conclusions. It was evident, however, that the volume of cream in bottled milk was influenced by various factors, some of which are apparently little understood. These factors include:

(1) The temperature to which the milk is heated.

(2) The length of time for which milk is held at the high temperature.

(3) The temperature of the heating medium with which the milk comes in contact during the heating process.

(4) The clarification of the milk.

(5) The type of apparatus used in treating the milk.

(6) The amount of agitation to which the milk is subjected, especially while hot.

This last factor has a greater influence upon

the cream line than is ordinarily appreciated. It has been stated to the writer, that in one instance where the cream line was materially reduced during the pasteurization process, the experiment was made of reducing the speed of the agitator which kept the hot milk in motion. After the speed had been so reduced it was found that the cream line upon the milk resumed its normal character.

It will also be noted from the foregoing description of the various methods of handling milk that in those plants where there was most agitation of the milk, and especially where the pumping of hot milk was practiced, the most difficulty with the cream line was experienced. There are possibly other factors which affect the cream line, among which may be:

1st, the age of the milk before pasteurization.

2nd, the grade of cows from which the milk was produced.

3rd, the fact that milk has or has not been frozen before being treated.

The tests made proved that it is entirely possible, under the most favorable conditions, to heat milk to 145° and hold it for thirty minutes without producing any injurious effect upon the cream line.

It is also doubtless true, however, that with some types of apparatus in use by the milk dealers and with some conditions under which milk is handled, a real difficulty is experienced in obtaining satisfactory results.

The chart shown in Fig. 34, drawn by Dr. North, indicates graphically the temperatures and holding times within which it is safe to heat milk without affecting the cream line. Such a chart cannot show the other factors which also affect the results, such as the amount of agitation, etc.

If the milk is being heated and an attempt is made to keep the milk at a temperature no lower than 145° F; it is almost certain that the temperature will at times reach from 146 to 148° or higher, and that it will at times be held longer than thirty minutes. It is almost mechanically impossible to so control the temperature of milk while heating that there will not be considerable variation.

The agitation probably has the effect of breaking up the fat globules into smaller masses. Since each globule of fat is surrounded with a film of casein or skim milk, which is heavier than the fat, it is evident that as the mass of fat becomes smaller, the proportional amount of the surrounding skim

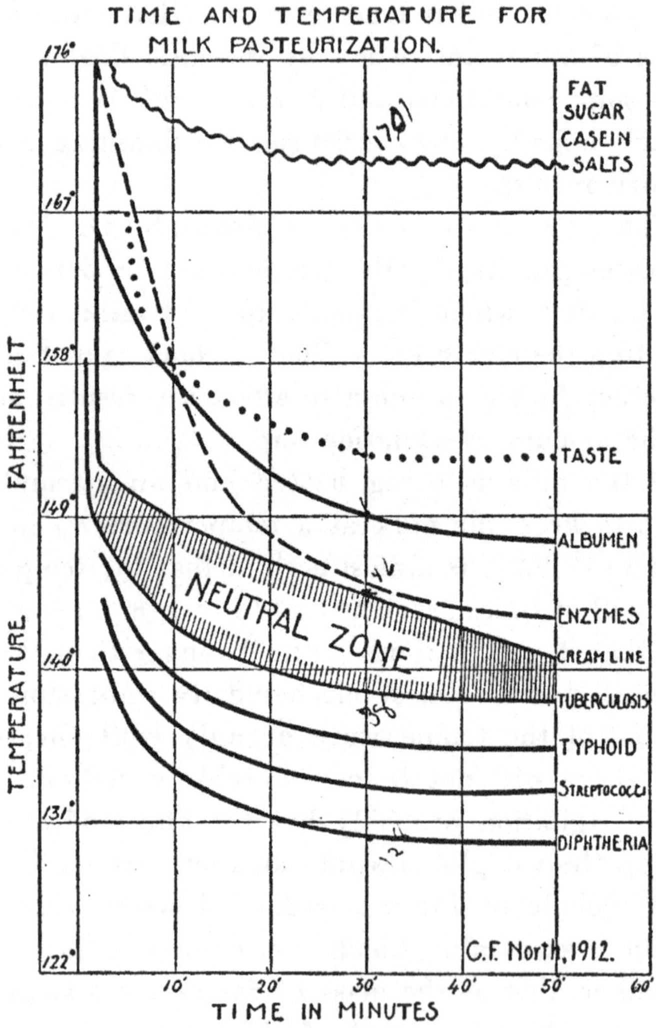
TIME AND TEMPERATURE FOR
MILK PASTEURIZATION.
176°
167°
158°
149°
140°
131°
122°
FAHRENHEIT
TEMPERATURE
FAT
SUGAR
CASEIN
SALTS
TASTE
ALBUMEN
ENZYMES
CREAM LINE
TUBERCULOSIS
TYPHOID
STREPTOCOCCI
DIPHTHERIA
NEUTRAL ZONE
C.F. North, 1912.
10'
20'
30'
40'
50'
60'
TIME IN MINUTES

FIG. 34

milk becomes greater, till we reach a point when the buoyant fat cannot lift the heavier envelope, and our balloon of fat stays down.

Just why the high heat and the long holding scatters the fat permanently through the body of the milk is not so clear. It may be due to chemical changes which occur.

In the pasteurizing equipment described on pages 124 and 131 it is claimed that the superheating of the held milk has a beneficial effect upon the cream lines.

The tables following show graphically the results of the tests made. They indicate the varied conditions under which milk was handled in the various tests and the percentage of the cream upon each sample as well as the percentage of butter fat which was determined by the Babcock test.

Test No. 1

Type of Pasteurizer:	Fig. 13 Disk Cooler			
Amount of Pumping and Agitation:	Pumped from ground floor to tank on 2nd floor. Flows to the **De Laval Clarifier.** Flows to pasteurizing tanks, to cooler, and to bottle filler.			
Sample Number:	1.	2.	3.	4.
Character of Milk:	Pasteurized.	Pasteurized.	Pasteurized.	Pasteurized.
Temp. to which heated:	142°	146°	146°	146°
Time for which held:	30 m.	34½ m.	83 m.	30 m.
Temp. of heating medium:	165°	165°	165°	165°
No. of hours sample was set:	4 hrs.	4 hrs.	4 hrs.	4 hrs.
Remarks:			Taken from the last milk leaving the tank from which No. 2 was taken.	
Percentage of Fat:	Not Determined	Not Determined	Not Determined	Not Determined
Percentage of Cream by volume:	15%	13%	8%	15%

Test No. 2

Type of Pasteurizer:	Fig. 13		Disk Cooler	
Amount of Pumping and Agitation:	Pumped from ground floor to tank on 2nd floor. Flows to the De Laval Clarifier. Flows to pasteurizing tanks, to cooler and to bottles.			
Sample Number.	1.	2.	3.	4.
Character of Milk:	Raw.	Raw.	Pasteurized.	Pasteurized.
Temp. to which heated:			145°	145°
Time for which held:			33 m.	30 m.
Temp. of heating medium:			180°	180°
No. of hours sample was set:	Abt. 11 hrs.	10 hrs. 45 m.	10 hrs. 17 m.	9 hrs. 3 m.
Remarks:	Taken before being clarified.	Taken after being clarified.	Taken from first milk leaving tank. Temp. 143½°	Last milk leaving tank from which No. 3 was taken. It had been cooled from 143½° to 62° in 40 m.
Percentage of Fat by Babcock test:	3.8	3.8	3.8	3.8
Percentage of Cream by volume:	19%	17%	12%	15%

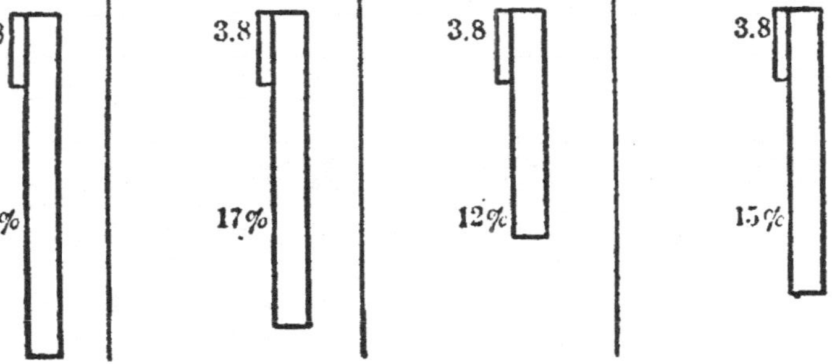

Test No. 3

Type of Pasteurizer:	Fig. 13		Disk Cooler	
Amount of Pumping and Agitation:	Dumped to weigh vat. Flows to mix vat. Pumped to tank 2nd floor. Flows to clarifier, to storage tank, to past tanks. Pumped to cooler.			
Sample Number:	1.	2.	3.	4.
Character of Milk:	Pasteurized.	Raw.	Pasteurized.	Pasteurized.
Temp. to which heated:	141°		145°	145°
Time for which held.	60 m.		60 m.	120 m.
Temp. of heating medium:	156°		158°	158°
No. of hours sample was set:	4 hrs. 6 m.	3 hrs. 20 m.	3 hrs. 40 m.	45 m.
Remarks:	Heating was slowly done on account of low steam pressure.	Milk had passed through clarifier.		Taken from same tank as that from which No. 3 was taken.
Percentage of Fat by Babcock test. Percentage of Cream by volume:	15%	15%	No Cream	No Cream

Test No. 4

Type of Pasteurizer	Fig. 5 and Fig. 22			
Amount of Pumping and Agitation:	Dumped into mixing vat, ground floor; pumped to tank on second floor; flows to clarifier, to mixing vat, to heater, to holder, to cooler.			
Sample Number:	1.	2.	3.	4.
Character of Milk:	Raw.	Pasteurized.	Pasteurized.	Pasteurized.
Temperature to which heated:		145°	144°	142°
Time for which held.		24 m.	24 m.	24 m.
Temperature of heating medium:		159°	157°	155°
Time of sampling:	7:15 P. M.	7:40 P. M.	8:55 P. M.	9:20 P. M.
Time of cream reading:	11:45 P. M.	11:45 P. M.	11:45 P. M.	11:45 P. M.
No. of hours set:	4 hrs. 30 m.	4 hrs. 5 m.	2 hrs. 50 m.	3 hrs. 25 m.
Remarks:	Sample taken from milk before clarifying.	Milk is probably held for longer than 24 minutes but the exact time could not be determined.	Milk is probably held for longer than 24 minutes but the exact time could not be determined.	Milk is probably held for longer than 24 minutes but the exact time could not be determined.
Percentage of Fat by Babcock test:	3.9%	3.9%	3.9%	3.9%
Percentage of Cream by volume	15%	10½%	8%	8%

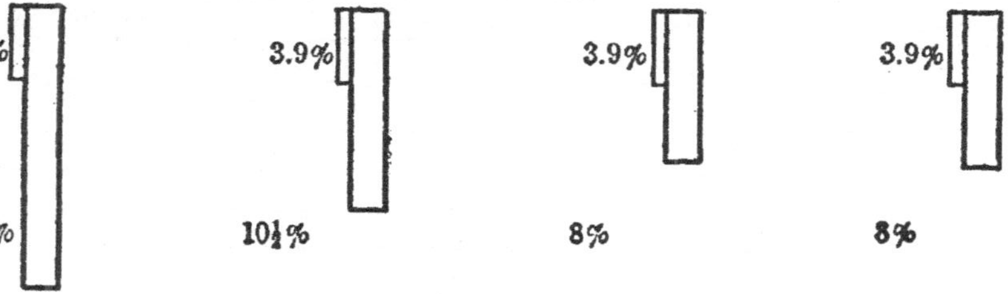

Test No. 5

Type of Pasteurizer	Fig. 4 (*modified*) and Fig. 21			
Amount of Pumping and Agitation:	Dumped into vat, flows to clarifier on lower floor, forced by clarifier to storage vat on second floor, flows to heater, having agitator, to holder where slowly stirred, flows to cooler and bottle filler.			
Sample Number:	1.	2.	3.	4.
Character of Milk:	Pasteurized.	Pasteurized.	Pasteurized.	Pasteurized.
Temperature to which heated:	143°	146°	145°	146°
Time for which held:	30 m.	30 m.	30 m.	30 m.
Temperature of heating medium:	165° to 209°	165° to 210°	165° to 210°	165° to 210°
Time of sampling:	10:05 A. M.	10:28 A. M.	10:45 A. M.	11:28 A. M.
Time of cream reading:	1:30 P. M.	1:30 P. M.	1:30 P. M.	1:30 P. M.
No. of hours set:	3 hrs. 25 m.	3 hrs. 2 m.	2 hrs. 45 m.	2 hrs. 2 m.
Remarks:	Recording chart indicated that milk was heated to from 140° to 146° It was not preheated before clarifying and pasteurizing.	Temperature chart showed milk held to from 142 to 148°. It was not preheated before clarifying and pasteurizing.	Temperature of raw milk 47° It was not preheated before clarifying and pasteurizing.	Milk was heated to 60° before clarifying and pasteurizing.
Percentage of Fat by Babcock test:	3.7%	3.8%	3.7%	3.9%
Percentage of Cream by volume:	11%	4%	7%	3%

Test No. 6

Type of Pasteurizer	Fig. 11 and Fig. 21			
Amount of Pumping and Agitation	Pumped from ground floor to mixing vat on fourth floor to clarifier to mixing vat; pumped to heater and holder where slowly stirred, pumped through cooler; flows to bottle filler.			
Sample Number:	1.	2.	3.	4.
Character of Milk:	Raw.	Pasteurized.	Pasteurized.	Pasteurized.
Temperature to which heated:		146°	143°	146°
Time for which held:		30 m.	30 m.	30 m.
Temp. of heating medium:		144° to 160°	144° to 160°	144° to 160°
Time of sampling:	8:30 A. M.	9:45 A. M.	9:50 A. M.	10:50 A. M.
Time of cream reading:	1:40 P. M.	1:40 P. M.	1:40 P. M.	1:40 P. M.
No. of hours set:	5 hrs. 10 m.	3 hrs. 55 m.	3 hrs. 50 m.	2 hrs. 50 m.
Remarks:				
Percentage of Fat by Babcock test:	3.4% 	3.5% 	3.75% 	3.35% 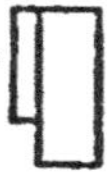
Percentage of Cream by volume:	13½%	No Cream Visible.	10%	5%

Test No. 7

Type of Pasteurizer	Fig. 9 and 3 tank holder, Fig 24			
Amount of Pumping and Agitation:	Dumped to dump tank, flows to mixing vat, pumped through heater and to holder, flows to cooler and bottle niler.			
Sample Number:	1.	2.	3.	4.
Character of Milk:	Raw.	Pasteurized.	Pasteurized.	Pasteurized.
Temperature to which heated:		142°	145°	145°
Time for which held:		Abt. 35 m.	30 m.	60 m.
Temperature of heating medium:		165 to 170°	165 to 170°	165 to 170°
Time of sampling:	9:25 A. M.	9:35 A. M.	10:04 A. M.	10:36 A. M.
Time of cream reading	12:40 P. M.	12:40 P. M.	12:40 P. M.	12:40 P. M.
No. of hours set:	3 hrs. 15 m.	3 hrs. 5 m.	2 hrs. 36 m.	2 hrs. 4 m.
Remarks:		The holder is a continuous type and holding time is not accurately determined.		
Percentage of Fat by Babcock test:	3.85%	3.8%	3.85%	3.7%
Percentage of Cream by volume:	14%	14%	14%	9%

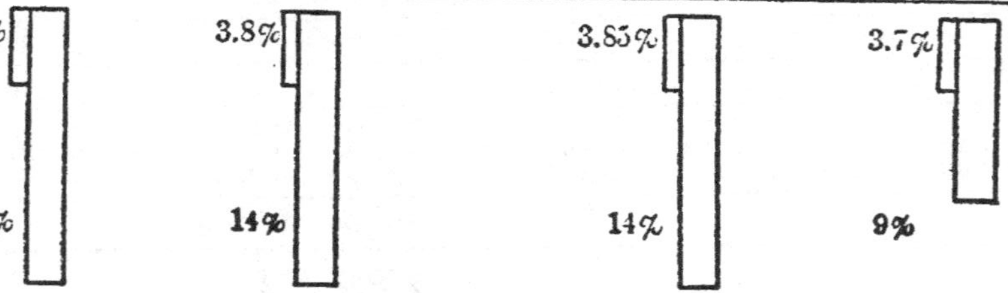

Test No. 8

ʼig. 7, and Fig. 27

clarifier
oler and

Remarks:	Sample taken from tank after milk was clarified.	Sample taken from dumping tank before clarification.	The holder is of the continuous type and the exact holding time cannot be accurately determined.	
Percentage of Fat by Babcock test:	3.0% 	3.5%	3.7% 	3.6% 
Percentage of Cream by volume:	12%	15½%	8%	7%

CONCLUSION

In conclusion, it may not be amiss to say that all the indications lead to the conviction that the pasteurization of milk is not, as some have claimed, a fad. It is coming more and more into use, as a recognized sanitary measure, which is at present necessary. Whether a time may come when it will be safe for large communities to consume their milk unheated is a question which at present is not a subject for practical consideration. The South and West, which have been slow to recognize the value of pasteurization, are coming into line, and the teaching of practical methods of handling apparatus used in pasteurizing milk is a wise thing for the dairy schools to take up. Practical methods of controlling the actual operation of pasteurizing plants should be taught, and instruction should be available for inspectors employed by municipalities, as well as for those who are to become such inspectors, in order that they may be equipped to render the most useful service to the communities when they assume their duties.

Other methods may be discovered by which milk may be rendered safe with less trouble and expense, but none such are now in sight, and until they appear our towns and cities must depend upon the proper application of heat to render milk safe as food.

It is due to the citizens that the actual treatment of the milk be so intelligently supervised and controlled that it will in reality be as safe as the people have a right to expect.

INDEX

A

B

C

INDEX

INDEX

H

I

INDEX

INDEX

R

S

Spotted something that isn't quite right?

Let us know and we'll fix it for future reprints. And as a thank you for your diligence, we'll send you a free Sothis book of your choice.

publisher@sothis.press

This page intentionally left blank

Except of course for that line above. And this one.
And whilst we have your attention, perhaps you'd
like to follow us on Instagram @sothis.press. Go on.

Made in the USA
Monee, IL
08 November 2026

68089576R00152